QH

Life
Manipulation

Life
Manipulation

FROM TEST-TUBE BABIES TO AGING

David G. Lygre

WALKER AND COMPANY
NEW YORK

The following excerpts, referenced in the Notes, are reprinted by permission of the Publishers:

From *The Biocrats* by Gerald Leach. Copyright © 1970 by Gerald Leach Productions Ltd. Published by McGraw-Hill Book Company. All rights reserved. Used with permission of McGraw-Hill Book Company.

From *The Biological Time Bomb* by Gordon Rattray Taylor. Copyright © 1968 by Gordon Rattray Taylor. Reprinted by arrangement with The New American Library, Inc., New York, N.Y.

From *The Mind of Man: An Investigation into Current Research on the Brain and Human Nature* by Nigel Calder. Copyright © 1970 by Nigel Calder. Used with permission of Viking Penguin Inc.

Extracts reprinted from *The People's Almanac*, by D. Wallechinsky and I. Wallace. Copyright © 1975 by David Wallace and Irving Wallace. Used by permission of Doubleday & Company, Inc.

Quotations from their works are reprinted by permission of the authors: Robert T. Francoeur, Herman Kahn, Robert L. Sinsheimer.

First published in the United States of America in 1979 by the Walker Publishing Company, Inc.

Published simultaneously in Canada by Beaverbooks, Limited, Pickering, Ontario.

ISBN: 0-8027-0632-0

Library of Congress Catalog Card Number: 79-63638

Printed in the United States of America

10 9 8 7 6 5 4 3 2 1

To Laurae

Contents

Acknowledgments

I would like to thank the taxpayers of Washington State, who paid for the sabbatical leave that made this book possible. And I am indebted to the scientists at the University of York (John Currey, Mark Williamson, Henry Leese, Simon Hardy, Peter Hogarth, John Sparrow, Alistair Fitter, Kim Booth, and others) and the University of Oxford (Christopher Graham, Bill Wood) who helped me during my stay in England.

I thank Stanley Falkow (University of Washington), Robert Lapen, Terry Devietti (Central Washington University), Gordon Prichett (Hamilton College), Mike Gray, and the Embree family (Erma, June, and Russ) for reading all or part of the manuscript on short notice and offering helpful suggestions. David Lundy, director of the Student Health Center at Central Washington University, sent me many useful medical references. I am also grateful to Richard Winslow of Walker and Company for his constructive criticisms.

To these people goes much credit for the content of this book. To me, however, must go full responsibility for the errors, personal opinions, and royalties.

Last, but most important, I thank my immediate family (Laurae, Jedd, and Lindsay) and my parents (Esther and Gerald) for their constant support and encouragement.

Preface

Man is a curious animal. He reaches out to life, to hold it in his arms, to stroke, taste, shake, sniff, and hear it. But he does much more. He thinks about what he has learned, he tells others, and he tries to understand better.

He has come a long way. No longer is he a captive of his physical world, for he is discovering how to mold his environment to himself. And now his search for understanding has brought him to the ultimate puzzle—himself.

His journey has been slow, but the pace is quickening. Man has invented powerful tools to probe himself, to unlock his intimate secrets. The story is just beginning.

Man's flight into inner space is revealing an exquisite machine. And the more he learns about that machine, the more he can tinker with it. But he also knows he is far more than a physical being; he is an emotional, ethical, social, and spiritual creature as well. These qualities will guide him as he decides when and how to intervene in the mechanics of life.

We can only hope he will choose wisely.

Reshaping life! People who can say that have never understood a thing about life—they have never felt its breath, its heartbeat—however much they have seen or done. They look on it as a lump of raw material that needs to be processed by them, to be ennobled by their touch. But life is never a material, a substance to be molded. If you want to know, life is the principle of self-renewal, it is constantly renewing and remaking and changing and transfiguring itself.

Doctor Zhivago by Boris Pasternak

I felt slightly queasy when at lunch Francis winged into the Eagle to tell everyone within hearing distance that we had found the secret of life.

The Double Helix by James Watson

1

Secrets

Reverence for life. Everyone I know believes in that. There's a mystery and wonder in life that fills us with awe and humility; it seems impossible that we could ever understand it all.

And yet, as we learn more about the mechanics of life, some of its secrets are disappearing. We're discovering how living creatures obey the same laws of nature as the rest of creation. Just as we are seizing control over our physical world, we are learning how to manipulate our biological world—to make "test-tube babies," to change our genetic makeup, to invent artificial body parts, to alter our brains, and to live longer. Indeed, we may even discover we can make life itself.

The revolution in biology could change not only our physical selves, but also the way we think of ourselves and others. The implications are truly stunning. As we put life under the microscope, carefully dissecting and analyzing each of its parts in the most exquisite detail, we may come to view life as a material, a lump of clay we can mold to our design. No longer will we accept the notion that a disease is incurable; no longer will we believe our bodies must eventually wear out and die; no longer will we accept whatever genetic features we happen to be born with. Indeed, we will no longer accept nature's way as inevitable, for we will wrest that power—and responsibility—from her.

Our technology promises to sweep us into a golden age, an age where we

1

will control the mechanics of life. But our glittering new tools are only part of the story, for they are intertwined with the whole fabric of our society. Indeed, they create social, ethical, and legal dilemmas we have not faced before, problems we must resolve if our new age is to be a better one. For example, we have learned how to separate procreation from intercourse; now we can consider them separate matters as we seek moral guidelines for our behavior. With artificial insemination and "test-tube babies" at our disposal, we no longer must accept infertility as nature's final verdict; now we must decide how and when we should take matters into our own hands. When human cloning becomes a reality, we will have to decide when, if ever, to use it. As we learn how to diagnose more genetic diseases early in pregnancy, we will increasingly base our abortion decisions on the quality of our fetuses, including their sex. And as we learn how to control our genetic makeup directly, we will have even more power to choose what kinds of people we want. Who will we accept as "healthy" and "normal"? And since we will be able to keep people mechanically alive almost indefinitely, we will also have to decide when to let them die.

These and other issues we will explore spring directly from our growing ability to manipulate life. The intrusion of this technology on our most intimate activities—having children, growing old, dying—forces us to respond. But what should we say? Should we say we will use these tools to benefit the largest number of people, or should our first priority be the dignity of each person? Whose rights are paramount—those of society as a whole, the prospective parents, the individual?

As the biorevolution speeds ahead, we find our legal, ethical, and social values lagging behind. It is hardly surprising. We pass new laws, for example, because of real problems, not hypothetical, pie-in-the-sky ones. The trouble is that some of these problems are not just hypothetical anymore. Our moral and social values also take time to change. Yet some of the questions we again face (What is life? When does life begin? When does it end? What is "meaningful" life? When should we "play God"?) are timeless, and it may be too much to expect that we will reach a consensus. Moreover, it is not at all clear how much our values should change just to accommodate our advancing technology. Indeed, science tells us what we *can* do; it does not tell us what we *should* do.

As scientists uncover the secrets of life, we can begin to glimpse its innermost nature. For most of our history life seemed an elusive butterfly that fluttered just beyond our grasp. Somehow, it seemed to flow into cells and organisms as if by magic. We didn't know exactly what it was. We knew it could be passed on from one generation to the next, but we didn't know any other way it could come into existence. For many centuries man could only engage in armchair speculation about these secrets. But then he discovered a new tool: the scientific method.

Louis Pasteur showed the power of this tool in 1864. The controversy was raging over whether life could be generated from nonlife, so the French Academy offered a reward to anyone who could settle the issue for once and for all. Pasteur pocketed the prize with a simple experiment. He put a sterilized broth in several flasks, some of which were open to microbes in the air. When microbes grew only in the flasks that could be contaminated, Pasteur concluded that life comes only from life. He declared: "Never will the doctrine of spontaneous generation arise from this mortal blow."[1]

He should have returned the prize. Pasteur's experiment swept away some false notions about the spontaneous generation of life, but that is not the same as disproving the idea itself. Indeed, this concept is still with us, though in more sophisticated forms. The reason is simple: Unless the earth has always had life, and that is most unlikely, life must have appeared here after a period of nonlife.

How did it happen? As scientists learned more about the architecture of life, they discovered that the same physical and chemical laws of nature operated in organisms as in everything else. This suggested that life is largely mechanical, so perhaps it could be made by natural events. They devised the theory of chemical evolution, which holds that the building blocks of life arose naturally on the primordial earth and then gradually formed more complex structures; from this, life emerged. Yet this idea clashed with another version: the belief that God blew the breath of life into the first organisms, and only they can pass this supernatural gift on to others.

Now we are learning new ways to pass life on. Indeed, our efforts to understand the mechanics of reproduction have gone forward at a startling pace. Two centuries ago we learned how to propagate human life by an "unnatural" method—artificial insemination. Now we can manipulate not only the sperm, but also the eggs, and produce "test-tube babies." We're also learning how to transplant embryos, and we can even analyze them before we implant them to see if they have the "right" genetic features. And looming on the horizon is a method for making as many genetic copies of a person as we want.

One of the crucial secrets of life was revealed in 1953 when James Watson, Francis Crick (both at Cambridge University), and Maurice Wilkens (at nearby King's College of London) discovered the structure of DNA, the hereditary stuff of life. They laid bare not only the architecture of DNA but also the way DNA furnishes the genetic information to our cells and to the cells of our descendants. In short, they uncovered the blueprint for living organisms.

With the blueprint in our hands, it seemed likely we could copy it and even change it, tailoring life to our desires. And, indeed, scientists have taken some dramatic strides in that direction. In 1973, for example, a group of California researchers devised a way to splice genes—bits of DNA made from simpler building blocks called nucleotides—from one organism into another.[2] And in 1976 H. G. Khorana and his co-workers at the Massachusetts Institute of Technology synthesized a gene 200 nucleotides long and inserted it into a living cell, where it functioned along with the cell's own DNA.[3] Then, in 1977, a research group at Cambridge University determined the complete nucleotide sequence for the entire DNA of a virus.[4] Since that sequence—5,375 nucleotides long—also codes for all the proteins in that virus, and the virus consists only of DNA and protein, one science correspondent proclaimed: "The breakthrough means that it is now possible to write down an exact chemical formula for a complete living organism."[5]

What are we going to do with all this knowledge? In what situations should we tinker with the mechanics of life? As we learn more about human genetics, for example, we are increasingly able to inform prospective parents about their risks of having defective children; we also can test their babies, and even their fetuses, for genetic disorders; and we are developing better methods to treat some of those disorders. But this brings problems as well. For example, the prospective parents will have to decide whether their risk of having a baby with a particular defect is too great to proceed with a pregnancy. And if a couple learns their fetus has a disorder, they will have to decide whether it is serious enough to warrant an abortion. Another problem, as we will see later, is that a long-term effect of these individual decisions is likely to be a deterioration of our gene pool; that is, in future generations more babies may be born with more defective genes. How shall we deal with this?

As we conquer still more diseases, we will be able to live longer. We may even uncover the secrets of aging and learn how to slow down its relentless toll on our bodies. And, in the meantime, we are learning how to replace many of our worn-out or diseased body parts with artificial parts or transplants. Yet these advances also carry a price tag: a rising population and an increasing proportion of old people within that population. How will we decide to allocate our limited resources? Who should receive the body parts that are in short supply? Can we afford research and expensive forms of therapy for rare diseases? At what point in a person's life, if any, should he not receive the most sophisticated technology modern medicine can provide?

While we struggle with such questions the biorevolution presses on, promising to make nature our servant. We will be able to control our evolution; we will decide what kinds of people to have; we will determine the right time to die. We may not reduce life to chemistry and physics, but our power over the mechanics of life will increasingly bring us to view life as an exquisite machine.

Will anything ultimately elude our quest? Will we be able to dissect out each piece of life, label it, and finally know there is nothing more? Will our very souls survive the search? According to Francis Crick: "I myself, like many scientists, believe that the soul is imaginary and that what we call our minds is simply a way of talking about the functions of our brains."[6] Indeed, scientists are probing even the timeless secrets of the brain; they are learning how we learn and remember; they are trying to make machines that can "think"; they are discovering how electrical, chemical, and surgical manipulations of the brain affect the way we behave.

Yet our tools for manipulating life need not rob us of our souls. In fact it is our ethical, social, and spiritual values—not our physical tools—that must command our future. We are becoming the custodians of an awesome control over life, and we must manage our tools wisely. We cannot ignore them, hoping they will go away. Nor can we pretend they lie far in the future and that there is no immediate need to deal with them. Indeed, the biorevolution is at our doorstep. Now let us explore those tools, their implications, and our response to them.

As a distinguished specialist in New York put it recently, people greeted artificial insemination first with horror, then rejection, then curiosity, followed by more study and finally by acceptance.

Joseph Fletcher

The great majority of people consider artificial insemination with the sperm of an anonymous donor appalling.

Bernard Häring

2

Artificial Insemination

Even if we discover we can synthesize life, this wouldn't be a practical way to produce simple organisms, much less people. Besides, we have a better method already. As Marilyn Monroe once said, "What do I think about sex? Oh, I think it's here to stay."[1]

Indeed it is. As far as I know, there hasn't been any public outcry against using sexual intercourse to reproduce. In fact, it is the only method worth trying even when it does not produce babies. Yet for many people, that is precisely the problem: This method does not work for every couple who wants children.

One couple I know decided not to start their family right away. They wanted to finish college first. After they graduated, he decided he needed more schooling and she wanted to test her wings in a new career; they could postpone having a family a little longer. When he finished graduate school, they moved to another city, where they both had good jobs. Just a couple more years—so they could get established and build up a little nest egg—and then they would start their family.

They were approaching their thirties now, and they realized they could not wait much longer if they wanted to be young enough to enjoy all those wonderful things they were going to do with their children. So they tried to have children. But nothing happened. After almost a year of disappointments, they began to study the problem. There was so much written about contraception, and so little about infertility. They did learn that fertility declines as people get older and their chances of an abnormal baby increase. Alarmed, they decided to see a physician. When he tested several semen samples from the husband, he found the sperm count in the normal range; the sperm had a normal appearance and movement.

She searched her past for a clue. She had used the pill for several years and then changed to an intrauterine device (IUD) when she heard about some dangers with the pill. Since an IUD can irritate the uterus and cause damage to the oviducts (Fallopian tubes), she thought her oviducts might be blocked. But when the physician tested her for this, the results were negative.

Since none of the tests revealed a problem, they thought maybe everything was normal and they were just having bad luck. So they tried harder. She charted her menstrual cycle faithfully, recording her temperature each morning. They learned to recognize when she was ovulating, for she had a regular pattern. They planned intercourse according to that chart and followed their schedule faithfully, even when they did not really feel like it. Still, nothing happened; each month brought a fresh disappointment.

Soon they were in their early thirties and they were beginning to despair. Maybe they should have had more thorough tests earlier, but they never expected it would come to this. Now it seemed a little late to go through more tests, for they probably would not do any more good than the others. Besides, even if their problem were solved, they would be fifty when their children were still growing up. So they decided to keep trying a while longer but also see whether they could adopt a child. Although they were told the wait would be at least two years, they applied.

Now they are in their midthirties. They are still waiting, and hoping.

As people postpone the start of their families, as they decide to have smaller families, and as abortion becomes more common, more and more couples are facing the prospect of a life without children of their own. And the increasing acceptability of contraception, abortion, and unmarried women keeping their babies has meant there are fewer babies available for them to adopt. As a result, many childless couples are looking to science for help.

That help is coming, for scientists are developing a variety of "artificial" methods for human reproduction. They are learning how to manipulate egg and sperm cells, and even embryos. As we will see in these three chapters, some fascinating doors are being pried open. In some cases we can only glimpse through the cracks, but already we can see applications we will applaud and others we will not. As we proceed, though, we will need to keep our eyes wide open to the potential benefits and hazards and to our limited power to prevent applications we dislike. For once we go through a door, we rarely choose to step back and close it.

LATEST TECHNIQUES

One door that has been open a long time is artificial insemination. Its first use in humans dates back at least to 1884, and probably to 1790.

In 1884, a wealthy businessman told his physician about a distressing problem: He and his wife had not been able to have a baby. Some tests revealed he was the likely cause of the infertility. When the physician discussed this case with a group of medical students, they persuaded him to try a semen sample from one of the most promising students. So when the businessman's wife came in for what she thought was a routine examination, she received a bonus—an anesthetic and an injection of semen.

It worked. After the child was born, however, the doctor decided the husband should know about the experiment. The businessman acccepted the news calmly but decided his wife should never know what had happened.

Now AI for humans is coming out of the closet. It has long been a standard technique in animal breeding programs; in fact, it is the preferred way to reproduce cattle and several other types of livestock. And AI is more common in human reproduction than most people realize. Official records aren't kept, but an estimated ten thousand to twenty thousand babies are born each year in the United States because of AI.[2] Worldwide, more than one million people may have been conceived in this way.[3]

The procedure is simple. First, the couple is examined, physically and psychologically, to determine whether some form of AI is appropriate for them. There are three categories: AIH, where the husband's sperm is used; AID, where the sperm donor is not the husband; and AIHD, where semen from the husband and another donor are mixed.

The semen usually is supplied by masturbation, though it also can be collected by condom during intercourse or by interrupting intercourse just before ejaculation. The physician usually uses the sample directly for insemination. But sometimes—especially with AIH—he uses a laboratory method to separate the semen into two fractions, one of which has a higher sperm density and motility. Another option he has is to freeze the semen for later use after mixing it with part of an egg yolk, antibiotics, and glycerol, a protective agent that resembles antifreeze.

The doctor has two common methods for insemination. Usually he uses a syringe to deposit the semen in the vagina. Another way is to put the sample in a small plastic cup, place it on the cervix, and leave it there for about twenty-four hours.

Timing is critical. As with natural insemination, the chances for pregnancy are best when the sperm enter about the time of ovulation. This is because of the short survival time for sperm (two or three days) and unfertilized eggs (about one day). So it is important to analyze the woman's menstrual cycle in order to pinpoint when she ovulates. Occasionally, she may need hormones to help regulate her cycle.

The routines vary, but doctors often perform AI at two-day intervals, starting just before the expected time of ovulation. There are typically two inseminations per cycle, though a larger number could be done with frozen sperm. They charge for each insemination is fifty to seventy-five dollars, with about half of the fee going to the "donor."

With AID, most women become pregnant within three to six months. If they do not, it is often because they have an infertility problem that had not been diagnosed earlier. AID with fresh semen produces about a 70-percent pregnancy rate; with frozen sperm it is about 50 percent.[4] The success rate with AIH, however, is about 20 percent or less.[5] But this lower rate is hardly surprising, for in couples seeking AI the husband usually has deficient semen.

APPLICATIONS

Most couples use AI because the husband is subfertile or infertile. But there are many other reasons for using this technique.

One possible use of AID is to avoid certain health risks. The main example would be a couple where the husband has a genetic disorder he could transmit to his children. Or he might be Rh-positive while his wife is Rh-negative and has been sensitized against carrying an Rh-positive fetus. In both cases AID with an appropriate donor would minimize the risks to their children.

Frozen semen invites some usual applications because the donor is not restricted by time or location. Fatherhood need not be interrupted by a husband traveling, or living apart from his wife, or even dying. Men killed in war have subsequently fathered children.

Another option is for men to provide ample samples for frozen storage and then have a vasectomy. This was a major reason for sperm banks springing into existence in the 1960s. But even though the price was moderate (today the annual charge is about forty to fifty dollars), sperm banks have not proved as popular as had been expected. One important drawback is that there's no assurance that the frozen samples will later produce children. For one thing, samples from different people vary considerably in how well they freeze; some may be good for ten years or more, while others deteriorate rapidly. Moreover, freezing and thawing generally lower sperm motility, and this decline may be especially severe in samples stored several years or longer. As one physician said, "You pays your money, and you takes your chances."[6]

Another possible problem is that sperm banks aren't tightly regulated, so their quality control could vary. Yet it would be rather important that the semen samples not be mislabeled, misplaced, inadvertently thawed, or accidentally used by someone else.

An unusual use of AI is for women to bear children for unmarried men, or for couples where the wife is infertile. Far-fetched? The following classified ad was placed in a California newspaper: "Childless husband with infertile wife wants test tube baby. English or Northwestern European background.

Indicate fee and age.'' He offered up to ten thousand dollars to anyone who would conceive ''his'' child by AI and then give him the child after birth. ''I'm a very moral man,'' he said. ''I don't want to meet the woman face to face, much less have sexual relations with her.''[7]

There have, in fact, been at least three documented cases of women bearing children for the biological father and his wife by AI[8]. And it is likely that there are other, undocumented cases where this has happened. In return for her services, and the baby, one woman received seven thousand dollars and all her maternity expenses.

Unmarried women—both heterosexual and homosexual—could use AID to bear children for themselves. Indeed, this has been done in several countries. These women see AID as a way to become mothers without having to resort to an affair with a man they do not intend to marry.

AI also offers the tantalizing prospect of preselecting the sex of a child. This is because the type of sperm that fertilizes the egg determines the child's sex. While each egg carries an X chromosome, a sperm carries either an X or a Y chromosome. If an X-carrying sperm fertilizes an egg, the child will be female (XX); the Y-bearing sperm will produce a male (XY).

Is it possible to control which type of sperm fertilizes an egg? It is becoming clear that the answer is yes. The reason is that the two types of sperm have differences (including size, shape, motility, behavior in an electric field, and sensitivity to acid) that can be used to separate them.

Landrum Shettles, a physician now in Randolph, Vermont, contends sex preselection will work even with intercourse. He has developed procedures to maximize the proportion of the desired sperm type trying to fertilize the egg. The most important factor is timing; intercourse two and a half or more days before ovulation favors girls, while a gap of one day or less favors boys. His recipes also include douches, controlling orgasm in the woman, and the position during intercourse. Shettles has reported that with these techniques, couples have children of the desired sex at least 80 percent of the time.[9]

But his prescriptions are controversial. A few physicians have confirmed that his system works; others are skeptical. One problem is that we need better statistical data for this method, not only on the sex ratios achieved but also on the total pregnancy rate and the health of the children born. Another complication is a report by Rodrigo Guerrero, a physician in Colombia. While he found that AI three or more days before ovulation produced mostly girls and insemination very close to the time of ovulation produced about 60 percent boys, he also reported that the pattern was reversed with natural insemination.[10]

Although sex selection may work with intercourse, it will be far more effective to separate the sperm types and then use AI. For one thing, more complete separations can be made in the laboratory than inside a woman's reproductive tract. And the pregnancy rate could be improved with AI by combining the treated samples to give high sperm counts before insemination.

It will work. Several separation methods have been tried in animal studies. AI with those samples gives a 60- to 80-percent success rate in producing the desired sex.[11]

Scientists also are learning how to separate the two types of human sperm. For example, the smaller, faster-swimming sperm that bear a Y chromosome are more likely to escape from certain barriers, or to swim over to materials on which they can be collected. One such method produced samples with 85 percent Y-type sperm; another gave 89 to 97 percent Y-type.[12]

The technology is advanced enough that several hospitals are trying this approach with couples who definitely want boys. According to a preliminary report, a medical team at Michael Reese Hospital in Chicago was able to isolate fractions with an average of 67 percent Y-type sperm. Because of a low sperm count, however, the pregnancy rate with those samples is lower than normal. Nevertheless, four of the first six conceptions have been males (67 percent); five of the babies were delivered at term, but one male fetus was spontaneously aborted.[13]

SOME CONCERNS

Safety

We might expect that laboratory manipulations would damage sperm and produce abnormal babies. But the data show otherwise. The evidence from animal breeding programs indicates AI is safe and effective. And although we have much less data for human AI, the conclusions are similar. According to one report, a study of twenty thousand children conceived by AI showed an incidence of abnormalities that was no higher than for children conceived by intercourse.[14] There are conflicting reports on whether AI slightly increases the rate of spontaneous abortions,[15] but the few studies with frozen sperm indicate a normal incidence of spontaneous abortions and defective children.[16]

Follow-up studies also indicate AI is safe. When fifty-four AID children were tested for physical and metal development, their only "abnormality" was their above-average IQ scores.[17] The effect on IQ is not surprising, though, for IQ intelligence is one of the criteria used in selecting sperm donors.

But that brings us to a safety problem—screening donors. One physician remarked: "Can I check the validity of their specimen, to make sure they didn't substitute someone else's? No. Can I check their medical history? No. Can I check to see if the person went out the night before delivering the specimen and got VD? No. I've just got to take his word for it."[18] In the real world, a thorough analysis of each donor would be prohibitively expensive, so his stated genetic history is usually accepted at face value. And it is cumbersome, and expensive, to test every sample from every donor for venereal diseases. Although there is a two-hour test that detects gonorrhea in most cases,[19] thus enabling the doctor to use the semen fresh, the usual practice is to check a donor's sample occasionally. This means there is some risk. Indeed, there has been at least one instance where gonorrhea was transmitted by AID.[20]

In practice, then, it is necessary to depend on the integrity of the donors and the people screening them. Most donors are interns, medical students, or other graduate students. This arrangement has worked reasonably well, but commercial pressures may develop for more donors. If that happens, the chance to supply semen for cash, trading stamps, or guitar lessons might attract a less lovely crew of donors. A few might even lie about their health if necessary. It would be like the problems with paid blood "donors."

Why would we want more donors? For one thing, there could be a closer match of the donor to the husband. And a wide selection might bring in more customers. Although features could not be guaranteed in the children, the discriminating shopper might be attracted to the catalog: "Donor six and a half feet tall, with high IQ, bulging biceps, music talent, blond hair, violet eyes, and whiter-than-white teeth."

Emotional Effects

Another concern is that AID (or AIHD) could disrupt family relationships. For example, the husband might feel sexually inferior; he may resent the donor and feel separated from his wife and "her" new child. And the wife may feel an attachment to the donor, and the baby growing inside her, but a detachment from her husband.

The problems are real. Consequently, a physician generally will not urge couples to try AID; the impetus must come from them. And before AID is done, there is typically an interview to assess whether both the husband and wife truly want AID and can cope with the psychological problems. Some couples are turned down.

Another safeguard is to select a donor who resembles the husband in race, skin shade, hair color and curl, blood type, and other features. In addition, the donor remains anonymous.

Because of these precautions, alienation is rarely a serious problem. To begin with, couples receiving AID tend to be thoughtful and responsible people with mature marriage relationships. And since AID, in contrast to adoption, lets the couple share the experiences of pregnancy and childbirth, the psychological effects are often positive. In a survey of 102 women who had tried AID (57 had become pregnant), 46 percent believed the experience had strengthened their marriage; the others reported that AID hadn't changed their relationship significantly.[21]

AIHD poses another set of problems, both physical and psychological. When two samples are mixed, the semen from one donor (usually the husband) may impair the other donor's sperm. Another problem is that couples clinging to the slight possibility that the husband would be the biological father may not be mentally prepared for AI involving another donor. So AIHD is rarely practiced. There is another tactic, though, that accomplishes the same thing: A couple could have intercourse the same day as AID.

A difficult issue parents face with AID (and AIHD) is whether to tell their child how he was conceived. The usual advice is that the couple not reveal this to the child, or anyone else. Since the donor is kept anonymous, there is no way for the child to learn who the genetic father is, anyway. But as AID becomes more common, both for married and unmarried women, keeping it a secret may become less popular.

There are few problems with AID children being accepted as part of the family. Indeed, one doctor has written: "These children mean more to families than children conceived in a normal manner. But for artificial insemination, motherhood would be denied the wife. Babies conceived in this manner are wanted children, often desperately wanted. I know of not a single case in my practice where things have worked out badly."[22]

Sex Preselection

One problem with using AI or intercourse for sex preselection is that they won't be 100 percent accurate. So there will be some unpleasant surprises. And when their expectations are not fulfilled, couples may feel more resentment and disappointment than they normally do when they get the "wrong" sex by chance. The child may suffer as a result.

But there is another side to this problem. Even now, without sex preselection, some children are psychologically abused because their parents wanted the other sex. If preselection methods worked most of the time, there would be fewer such children.

Another practical problem with sex preselection by AI is that it will not be cheap. Unless the government or health insurance companies underwrote the cost (ultimately sending us the bill), sex preselection would be an option mostly for the middle- and upper-income families.

Yet there are situations where sex preselection clearly would be beneficial. One use is to avoid having children with diseases such as hemophilia and Duchenne muscular dystrophy, which are carried on the X chromosome. Women who carry such a disorder do not have the disease if their other X chromosome is normal. But their sons will receive their only X chromosome from the egg, so there's a 50-percent chance that they would have the disease. Although half of their daughters (on the average) would be carriers, none would have the disease because they will have at least one normal X chromosome, the one supplied by the sperm. So if the wife were a carrier, the couple could avoid having a child with muscular dystrophy or hemophilia if they could preselect girls. But since no preselection method will be perfect, they would still have to consider a backup plan—diagnosing the sex of their fetus early in pregnancy and aborting in case of a male.

Sex preselection could also affect our population patterns. This prospect has been a great boon to demographers and sociologists, who have yet another topic for their questionnaires and speculations.

One effect would be on family size. Although some couples would decide to have another child if they could be sure of the sex, the larger effect would likely be in the opposite direction: Couples would have fewer children if they could get a child of the desired sex on each try.

Another effect might be an excess of boys. Polls and studies of family reproductive patterns have indicated a preference for having boys, especially as the first child. Amitai Etzioni, a sociologist at Columbia University, has speculated that if sex preselection produced an abundance of males, our society might have less culture, less religious activity, more aggression and crime, increased prostitution, increased homosexuality, and greater interracial and interclass tensions.[23] It is a fine commentary on the male species.

A survey of six thousand married women, however, indicates the results wouldn't be so drastic. Most parents want both boys and girls, with a son coming first. Parents who already have children say they would use preselection mostly to balance the sex ratio of their children. So we could expect an initial surplus of boys, followed by a wave of girls, with a fairly normal sex ratio in the end. And even those mild effects would depend on having a safe, inexpensive, and reliable method available, and large numbers of people choosing to use it. Yet according to the survey, many people would not use it at all.[24]

There is just one snag as far as the sex ratio is concerned. Suppose it turns out for a while, as is the case now, that our methods work better for preselecting boys.

LEGAL ISSUES

We will examine legal issues for many of the techniques described in this book. But first, we need to understand the role of law in our society. According to one authority, "the object of civil law is not to impose strict rules based on religious, political or racial assumptions but to try to derive empirically a set of standards to ensure the protection of people in both their individual lives and their social relations."[25]

It is important for us to realize the difference between legality and morality. The law may allow practices that individuals consider immoral, but the effect should be to grant each person the maximum freedom to exercise his own beliefs, so long as this does not harm others. For example, laws permitting abortion on demand—whatever their merits—allow individuals to act on the basis of their own moral code in this matter. Now we can question whether fetuses are persons and should be protected under the law (we will discuss this in the next chapter), but the point here is that neither a restrictive or permissive law on abortion would settle the moral issue. Each of us must reach our own verdict on that, regardless of the law.

Now let us turn to the legal concerns with AI. AIH has the fewest problems, though there is one minor issue. Suppose a child conceived by AIH (or even AID or AIHD) were born to a couple that hadn't had sexual intercourse. Would they still be entitled to have their marriage annulled on the grounds of nonconsummation? There was one such case in England involving AIH, and the annulment was granted.[26]

This situation, however, is more a legal curiosity than a serious problem, for couples usually use AI only after trying the real thing. Indeed, nonconsummation is not known to be a common practice among newlyweds.

AID and AIHD raise more substantive problems. In 1956 an American court stated: "Heterologous artificial insemination [AID] . . . with or without the consent of the husband is contrary to public policy and good morals, and constitutes adultery on the part of the mother. A child so conceived is not a child born in wedlock and therefore illegitimate. As such it is the child of the mother and the father has no right or interest in said child.[27] But most rulings have not concurred. Most courts have ruled that AID is not adultery because there is no sexual intercourse. And several have ruled that AID children are "legitimate."

Because of the hazy legal situation, some couples have taken steps to protect their AID children. One option is adoption, though this has the disadvantage that the parents might have to reveal how their child was conceived. More commonly, a couple may change doctors after AID and not disclose how conception occurred. The new physician, in good conscience, would list the husband's name on the birth certificate as the father. It is not exactly a novel idea; there have been one or two cases of adultery where similar deceptions were used.

A better solution is to make AID children legitimate, provided the insemination is performed by an authorized person and with the written consent of the husband and wife. Several states and countries have passed such legislation, and this ought to become our standard legal position.

Hiring surrogate or "host" mothers to bear children by AI poses several legal problems.[28] It is not clear that a woman can legally make money for those services. And although she may consent in writing to give up the baby to the couple, that agreement might not hold up in court if she changed her mind; in fact, she might even be able to sue the father for child support. The law is mostly silent on these issues.

The physician also is vulnerable, even in conventional AID. If he performs AID without the husband's consent, he may be liable for child support. And if the semen transmits a venereal or genetic disease, both the physician and donor might be liable.[29] This is perhaps the only exception to a policy of keeping the donor anonymous.

An anonymous donor, however, poses another problem—the possibility that two people sired by the same donor would happen to produce children. This inadvertent inbreeding would increase their risk of having children with genetic defects. One such marriage actually was planned and then cancelled when a physician revealed that the prospective bride and groom had the same biological father.[30]

But this danger can be kept at a minimum by limiting how many children each donor can produce by AID (perhaps five), and by using his samples only in diverse locations. With these precautions, according to one estimate, such marriages would occur in Britain only once every fifty to one hundred years.[31]

Indeed, adultery poses a greater risk of inadvertent incest. The rate of "illegitimate" births now exceeds 10 percent of all births and is as high as 50 percent in some urban areas.[32] For example, one modest study of an English town revealed the immodest fact that at least 30 percent of the husbands could not possibly have been the biological fathers of their children.[33]

ETHICAL ISSUES

AI raises several moral issues that also apply to other techniques we will discuss later. Here we will consider four of them.

Intervention in Nature

Some people contend AI is immoral because it is unnatural. For example, the Catholic Church has condemned one feature of AI—masturbation—on these (and other) grounds, or anywhere else.

To begin with, these objections rest on biblical interpretations that are arguable. Furthermore, the issue here is not masturbation in general, but its specific use to supply semen for AI, including AIH. If this is a serious problem for the couple, however, there is a cumbersome solution: The donor could have intercourse with his partner, and the semen could then be retrieved from a condom, or from her vagina.

Yet even if the semen were collected after intercourse, the recovery of sperm and the insemination itself could be considered unnatural. So here we need to examine a more basic question: Is something that is unnatural therefore immoral?

First, we must decide what we mean by "unnatural." That word generally means anything artificial, anything that violates natural law, or an abnormal pattern of behavior. But it is hard to apply that definition, for we are part of nature. When we develop new ways to manipulate our environment, is this "unnatural"? It it "unnatural" for us to use our intellect and skills to improve the quality of our lives, both physically and culturally? Hardly.

Then we come to the question of morality. We decided long ago not to endure the consequences of nature where we could intervene to reduce suffering. Indeed, this is the cornerstone of a civilized and humane society. In medicine, for example, we use a wide array of gadgets, surgical methods, and synthetic drugs. If we consider them "unnatural," and if we believe everything unnatural is immoral, we must condemn cesarean sections, bone marrow transplants, heart pacemakers, braces, glasses, kidney dialysis units, hearing aids, tooth fillings, false teeth, artificial hip pins, and many of our anesthetics, antibiotics, and anticancer drugs.

The crucial moral issue is *not* whether certain methods are "natural" or "unnatural"; it is to decide wisely when we should use them. Joseph Fletcher, an ethicist at the University of Virginia, has said: "Socrates thought it better to be an unhappy man than a happy pig. The pig's satisfaction with things as they are contrasts with a human being's struggle to make things better. Willingness to run the risks of requisite change and improvement is what makes human beings human. Humanness is courage married to reason."[34]

The Marriage Relationship

Some people object to all forms of AI (but especially AID) on the basis that God joined procreation with sexual intercourse, and he made them part of the marriage relationship; therefore, other methods of reproduction are immoral because they are outside God's plan. According to Paul Ramsey, an ethicist at Princeton University: "Since artificial insemination by means of semen from a non-husband donor (AID) puts completely asunder what God joined together, this proposed method . . . must be defined as an exercise of illicit dominion over man no less than would be the case if his free will were forced."[35]

Others, who are less certain of the details of God's plan, respond that the essential feature of marriage is the love between the husband and wife; intercourse and reproduction are important manifestations of that love, but they are secondary to the relationship itself. Although an AI child is not conceived by intercourse, and in AID someone besides the husband and wife is involved, the true marriage relationship is not necessarily violated. In fact, experience with AI shows that its major effect, if any, has been to strengthen that relationship. Furthermore, most couples seeking AI are not willfully separating intercourse from procreation; it is not their choice—indeed, it is their misfortune—that they are biologically denied the opportunity to have children by sexual intercourse with each other.

At least AID is far preferable, morally, to an Old Testament practice of adultery. According to *Genesis* 16:2: "Sarai said to Abram: 'The Lord has kept me from bearing children. Have intercourse, then, with my maid; perhaps I shall have sons through her.' Abram heeded Sarai's request."[36]

Adultery

Some people contend that in AID the woman is committing adultery with the donor. As we have seen, there is a legal precedent for this view.

Nevertheless, the charge hardly deserves comment. Adultery is commonly defined as "voluntary sexual intercourse between a married person and a partner other than the lawful husband or wife."[37] Yet in AID the donor does not meet the woman; in fact, he does not even know who she is. It is nonsense to equate sexual intercourse with the physical transfer of semen.

Some might argue, however, that in AID the woman is being unfaithful in the psychological sense. First of all, that is not adultery. Besides, this objection is groundless anyway, at least where the husband and wife both give voluntary, informed consent before AID is performed.

There is only one clear link with adultery: A woman could use AID to camouflage real adultery.

Overpopulation

Since overpopulation is a serious concern, we may object to any artificial method, including AI, for producing more people. Yet even though overpopulation is a major problem, it is hard to justify this as a sufficient reason to oppose AI.

First of all, we do not know what the long-term effect of AI will be. Currently, it is a minor factor in our population growth. But as sex preselection becomes an optional feature of AI, the net effect may well be to produce smaller families.

Second, until we require fertile couples to limit their family size, it is simply unfair to insist that, in the name of population control, infertile couples bear no children. On what basis could we justify setting a limit (zero) only for them? Because nature ordained it? If we believed that, we should also oppose the use of hormones and corrective surgery to overcome infertility. We might even accept obstetric problems as fate, and not intervene with fetal transfusions, cesarean sections, and the like. Instead, we could let the baby (and, sometimes, the mother) die in the name of population control, and in the name of bowing to nature.

No, thanks. We ought to intervene where we can to promote human happiness and dignity. The burden of overpopulation rests on all of our shoulders, and infertile couples who could bear children should not be singled out to carry an extra portion of the load. They, like all couples, should consider alternatives such as adoption or becoming foster parents. But they should have the same opportunities, and concomitant responsibilities, as other couples.

AN OVERVIEW

We have examined many aspects of AI. We can see problems and applications we dislike, to be sure. But we need to keep AI in perspective.

The major use of AI in humans is to help infertile couples have children. In the United States, as elsewhere, about 12 percent of all couples have an infertility problem, and the husband is responsible in about one-third of those cases.[38] Corrective surgery and hormone therapy will help a few of them. But AI, and especially AID, is the only way for most of those couples to have children, apart from adoption or having foster children. And adoption is becoming much more difficult.

For many couples, having children is one of the most rewarding experiences in life. That is what AI is about. AI is not for everyone, but as its use grows, and the problems are confronted honestly, "one solid fact is evident: artificial insemination is providing an increasing number of involuntarily childless wives with a means of satisfying the yearning for motherhood, and bringing joy to thousands of families."[41]

It is one thing to accept voluntarily the risk of a dangerous procedure for yourself. . . . It is quite a different thing deliberately to submit a hypothetical or unborn child to hazardous procedures which can in no way be considered therapeutic for him and are "therapeutic" for you only in that they "treat" your desires, albeit unobjectionable ones. . . . Morally, it is insufficient that your motives are good, that your ends are unobjectionable, that you do the procedure "lovingly," and even that you may be lucky in the result: You will be engaging nonetheless in an unethical experiment upon a human subject.

Leon Kass

Prediction for 1981–1990: A woman will be able to walk into a kind of "store" where frozen human embryos are sold in packets, and select a baby with specific physical, mental, and emotional attributes. Then, she will be able to have the embryo implanted in her uterus by her doctor and in 9 months will give birth to a baby. It will be possible for such embryos to be guaranteed free of genetic defects.

E. S. E. Hafez

3

Overcoming Infertility

AI isn't even half the story. Women have more infertility problems than men. And it is little wonder; in order to substitute for their reproductive role, we would need a bit more equipment than a syringe.

First, let us review how nature does it. Most women ovulate on the eleventh to the fifteenth day of the menstrual cycle. Each month one of their several hundred thousand eggs bursts out of a tiny blister on the surface of an ovary and makes its way through a fingerlike passageway into an oviduct. The egg spends about three days on its four-inch trip down the oviduct, but fertilization must occur the first day if the egg is to produce a baby.

After insemination, several hundred million sperm try to fight their way through mucus, acid, and other obstacles. Some make wrong turns at the uterus and go up the other oviduct. But in an hour or two the survivors begin arriving, having completed their seven-inch journey into the upper part of the oviduct. There the egg dances with many partners at the same time, each trying to penetrate her tough hide. When one finally succeeds, nature declares the egg off limits to the other sperm.

The fertilized egg begins to divide. A few days later the embryo (which will become a fetus in two months) has sixty-four or more cells. One week after fertilization the embryo begins to burrow into the uterine wall, establishing contact with the mother's blood system with special cells that will develop into a placenta. Implantation triggers the release of hormone messengers that pass the word: Until further notice, menstruation is canceled.

It is complicated enough that there are many places where something can go wrong. So now we are developing tools to assist nature when she is not able to be a mother.

NEW TECHNIQUES FOR "TEST-TUBE" FERTILIZATION AND EMBRYO IMPLANTS

In Vitro Fertilization and Embryo Implantation

Sperm can fertilize eggs not only in the Fallopian tubes, but also in test tubes. This is called in-vitro fertilization. (*In vitro* literally means "in glass," and it refers to something normally occurring in the body happening in an artificial environment.)

The first step is to collect the eggs. The ovaries are stingy with their monthly allotment, but hormone injections can persuade them to be more generous. This is called superovulation. Then with large animals, such as cattle and horses, the physician can insert a flushing device through the cervix and simply wash out the eggs (fertilized or unfertilized) into containers.[1] With humans (and smaller animals), however, the procedure is more complicated. About thirty-two hours after the patient receives her hormone injection, the physician makes a small incision in her abdomen and examines her ovaries with a laparoscope, a thin, telescopelike tube with an eyepiece and internal lighting. When he finds the telltale blister, he inserts a suction needle and removes the egg(s). The surgery is minor enough that the patient usually needs to spend just one night in the hospital to recover.

After waiting in a special medium for a few hours until they are ready to be fertilized, the eggs go into a suspension of sperm. Within several hours, one of the sperm pierces the outer layer and fertilizes the egg. These "test-tube" fertilizations have been done on eggs from many animals, including the rat, cat, monkey, mouse, and man.[2]

Still, it isn't as easy as it sounds. Sperm do not fertilize eggs in test tubes as efficiently as in the body. One problem is that until the sperm go through a chemical change, called capacitation, they cannot pierce the outer wall of an egg and fertilize it. In natural conceptions, this change occurs when the sperm swim through the uterus on their journey to the egg. But this does not happen during in-vitro fertilization experiments unless scientists "capacitate" the sperm by mixing them with materials from the uterus, or with certain artificial media.[3]

The fertilized eggs can develop at least several days outside the body. The embryos grow under special laboratory conditions until they reach the stage where they can attach to the uterus. Then it is time to try implanting them.

Implantation seems fairly simple, but timing is critical. One question is how long the embryo should first develop in the laboratory. With humans, the best time seems to be about two and a half days. It is also crucial that the recipient is at just the right stage in her hormone cycle to receive the embryo, so she may receive hormone treatments before and after the implantation attempt to prepare her uterus. There are two methods for the implantation itself. The physician may make an abdominal incision and place the embryo in her uterus. But the preferred method, which avoids surgery and requires no anesthesia, is to send the embryo through a soft, thin tube that passes through the vagina and cervix into the uterus.

These methods also work for removing and transplanting embryos. This is not exactly a new idea; the first successful embryo transfer was done with a rabbit in 1890.[4] With both cows and horses, scientists have used nonsurgical methods for collecting and implanting embryos and producing offspring.[5] And in 1975 scientists at Texas A & M University removed a 5-day-old embryo from one baboon and transplanted it into another. After a normal, 170-day pregnancy, the proud new mother had a healthy male baboon by cesarean section.[6]

We might expect that the new mother-to-be would reject her uninvited guest, just as she would reject other transplants. But experiments have shown otherwise. For many species, including the rat, mouse, rabbit, pig, sheep, and cow, the uterus of one female will accept an embryo from another animal of the same species. Indeed, the uterus is very tolerant of its tenants, and we should be grateful. For this also means a woman can choose her reproductive partner from a wide variety of males.

In fact, embryos can grow in some strange places. Mouse and other rodent embryos have developed awhile in the testis, kidney, brain, spleen, and anterior chamber of the eye.[7] Not only can embryos grow in the wrong organ and sex, but they can even develop in the wrong species. Sheep, cattle, and pig embryos live happily for several days inside the uteri of rabbits.

We are also learning how to freeze, store, and revive embryos. In one study, mouse embryos frozen up to eight days were thawed and implanted in mice; more than 40 percent of them developed into full-term fetuses or baby mice, apparently normal.[8] Indeed, scientists have produced baby mice from embryos that were frozen up to eight months, sent by air mail to another location, thawed, and then implanted.[9] Frozen embryos also have developed into calves.[10]

Could we produce humans by fertilizing eggs in a test tube and implanting the embryos? In the late 1960s a few physicians began trying to do it. Their patients were infertile women whose eggs were fertilized by sperm from their husbands. The results of their experiments were evident a few days after the implantation, when menstruation began.

Then the breakthrough came. Or did it? In July 1974 Douglas Bevis of the University of Leeds announced that three babies had been born after in-vitro fertilization and embryo implantation. Although he refused to identify the people involved, citing their need for privacy, he said two of the babies were in England, the other was in Italy, and all were apparently normal. Later, saying he was "fed up" with all the publicity, he announced his retirement from this type of research. But he has not published this work in a professional journal, or submitted other proof, so many scientists doubt that he really did it.[11]

In 1976 Patrick Steptoe (a gynecologist at the Oldham, England, and District General Hospital) and Robert Edwards (a physiologist at Cambridge University), pioneers in developing these methods for human embryos, reported a pregnancy that lasted about ten weeks. But the embryo implanted in the oviduct instead of the uterus. As with natural pregnancies of this type (ectopic pregnancies), the fetus could not develop normally and there was a miscarriage.[12]

Then it happened. After more than eighty unsuccessful attempts Steptoe and Edwards produced a full-term pregnancy. The key changes they made were to stop using superovulation and to have the embryo spend less time (two and a half days) developing in the laboratory, reaching only the eight-cell stage, before implanting it. The parents were Lesley Brown of Bristol and her husband, John, an employee of British Rail. She was infertile because of blocked oviducts, and surgery had failed to correct the problem. Indeed, Steptoe had found it necessary to remove her diseased oviducts, eliminating any chance that she could conceive in the usual way. But on July 25, 1978, Lesley gave birth to a five-pound, twelve-ounce baby girl by cesarean section. Hailed as the world's first "test-tube baby" (an exaggeration by the media), Louise was a healthy, normal baby.[13] Within the next three months came announcements of other such pregnancies in progress and a "test-tube baby" being born in India.[14]

So in-vitro fertilization and embryo implantation with humans are no longer science fiction. They are here now.

Artificial Wombs
Sometime in the twenty-first century we may learn how to develop embryos into babies without implanting them at all. Then we would truly have "test-tube babies."

It will not be easy. Fertilizing an egg and nourishing the tiny embryo for a few days in the laboratory is far simpler than growing the embryo into a baby. Before we could do that, we would have to unlock the mysteries of the uterus and discover all the ingredients that prescribe a healthy baby. And we would have to find a substitute for the fetal lifeline, the placenta, which flushes wastes from the fetus and lets in nutrients and oxygen from the mother's blood.

A complete artificial womb lies far in the future, but scientists are working on it. One way is to fertilize eggs in the laboratory and grow them there as long as possible. In 1959 Daniele Petrucci of the University of Bologna announced he had grown a human embryo for twenty-nine days but then terminated it because it was abnormal. A few months later he said a similar experiment had progressed well for fifty-nine days, until the fetus died because of a laboratory error. In 1966 a Russian research group that had consulted with Petrucci announced they had grown 250 human fetuses entirely in the laboratory, the oldest one living six months and reaching a weight of one pound, two ounces.[15] Scientists are skeptical of all these reports, however, because few details have appeared in scientific journals.

A less spectacular, but more realistic, way to develop an artificial womb is to interrupt pregnancies at different stages and try to grow the embryo or fetus in the laboratory. D. A. T. New and his coworkers at Cambridge University have removed mouse and rat embryos, complete with their placentas, one to two days after implantation and cultured them for three or four days. That may not seem like a long time, but it is 15 to 20 percent of the normal pregnancy period, and it is a time of enormous growth and development. Those fetuses in the laboratory developed beating hearts, small limb buds, and even, in a few cases, a functioning blood circulation. Then they died.[16]

One problem is finding a substitute for implantation. Scientists have tried attaching mouse embryos to eye lens tissue from cows, and to connective tissue from rat tails. In the latter study, done by Yu-chih Hsu of Johns Hopkins University, twenty-five embryos developed in the laboratory for ten to fourteen days, and seven of them reached the stage of having heart contractions. But they were all abnormal.[17]

Another problem is that the placenta does not fully develop under the laboratory conditions tried so far. Unless we can find those right conditions, we would need some other way to supply our "test-tube babies" with food and oxygen and cleanse them of toxic wastes.

One idea is to bathe them in nutrient broths inside high-pressure oxygen chambers. Rabbit and mouse fetuses have survived short periods inside such chambers, with their hearts beating as long as thirty hours.[18] But high oxygen pressure damages fetal tissues, so this is not a promising solution.

Another candidate is perfusion, a technique for supplying nutrients through tubes connected to blood vessels. The doctor can connect tubes to the fetus at the umbilical cord, which has two arteries and one vein. A pump circulates the blood out of the body, where it can be cleansed, oxygenated, and fortified with nutrients, and then returns it to the fetus. In a limited way, then, perfusion is a combination food source, artificial lung, artificial kidney, and artificial heart.

Perfusion works well—in theory. One drawback is that it cannot be used in the earliest stages of development, where there is no circulatory system. Yet even when the blood vessels develop, they are so tiny and fragile that it is hard to keep the blood circulating at an adequate rate. And the tubes often slip out. Another problem is that oxygenating devices often cause the blood to clot. We could solve that by using anticoagulants, but that would increase the risk of the fetus hemorrhaging.

Much of our knowledge about perfusion comes from experiments with sheep, which have a pregnancy period of 148 days. Scientists have perfused mature sheep fetuses (about 120 days old) as long as fifty-five hours.[19] And fetuses over 130 days old that were perfused up to twelve hours have survived to a successful "birth."[20]

Because of ethical restrictions, only a few perfusion studies have been done with human fetuses. Fetuses weighing less than one pound have shown movement and a normal heart rate for up to twelve hours of perfusion.[21] In another study, five premature fetuses in acute distress were perfused for up to three hours. All except one eventually died, and it wasn't clear whether perfusion had helped any of them.[22] A more promising case was a twenty-six-week fetus, weighing about two pounds, that was kept alive by perfusion for over five hours, in apparently good condition. Then a small tube slipped out and could not be reinserted, so the fetus died.[23]

APPLICATIONS

In-Vitro Fertilization and Embryo Implantation

In animal breeding programs, embryo transplants are a way to produce more offspring from the same pair of genetic parents. The method works like this: A prime cow superovulates and then receives semen from a prized bull by AI; her embryos—as many as fifteen or twenty—are removed, transported, and transplanted into cows of a less distinguished breed, which will bear the preferred type of calves. This enables stockmen to build herds of breeding stock more rapidly than by other methods. But embryo transfers are expensive enough (about 2,000 dollars per embryo, with about a 40-percent chance that an implanted embryo will produce a calf) that they are used only for valuable breeds such as Simmentals, superior beef cattle that cannot be imported in large numbers because of quarantine requirements.

There are several ways to store embryos traveling to other destinations. For short periods a traveler could simply carry in his pocket, close to his body, a test tube containing the embryos. Frozen embryos would be another solution, and they would make it easier to synchronize implantation with the recipient's hormone cycle, but this technology is not fully developed. Yet another option is to transfer embryos from large animals into the uteri of rabbits and ship the rabbits. With this method sheep embryos from England have become sheep as far away as South Africa and Japan.

What we can do with animals we presumably could do with people—if we wanted to. We would hardly choose to mass-produce children from the same set of parents, but we might find some other applications more attractive.

The main use would be to help infertile women. Many thousands of women in the United States alone are childless because eggs and sperm cannot move freely through their oviducts. The oviducts can be damaged in several ways, including infections, an ectopic pregnancy, a tubal ligation, or scar tissue forming around fragments that escape from the uterine lining. When blocked oviducts are the problem, couples still could have children by having the wife's eggs removed, fertilized in vitro, and then implanted in her. And with this method, couples could try for two (or even more) children from the same pregnancy. Indeed, multiple births might become a common way to make up for lost time.

Sometimes surgery also can help those couples. In one group of sixty-six patients who had reconstructive surgery on their oviducts, twenty-six became pregnant.[24] Another, more distant possibility is an oviduct transplant. Here the major obstacles are rejection and the difficulty of reconstructing the intricate network of blood vessels around the oviduct. But already one lucky rabbit has become pregnant after receiving a new oviduct and ovary.[25] A few women also have received oviduct transplants, but none has become pregnant yet.[26]

Women who ovulate irregularly, or not at all, could still have children by in-vitro fertilization and embryo implantation. Many of them, however, will find hormone therapy a better option. Indeed, this method of treatment has improved dramatically in the last few years, though there still is a risk of causing unwanted side effects and superovulation. Many pregnancies with six or more fetuses—and one with fifteen[27]—have followed hormone treatments. Another, unlikely option is to have an ovary transplant. Animal experiments indicate the ovaries are less vulnerable to rejection than most other transplanted tissues.[28] Women have received ovary transplants, and there is at least one report of a pregnancy after the operation.[29]

Fertilizing eggs in test tubes and implanting them opens up many possibilities, for the recipient and her partner would not necessarily supply either the egg or the sperm. So a wife could bear a child with any of the following combinations of genetic parents: husband-wife; nonhusband-wife; husband-nonwife; nonhusband-nonwife.

We can imagine situations where a couple might prefer to use eggs from someone else even if the wife could supply them. For example, if she might have Huntington's chorea, if she carried a genetic disease such as muscular dystrophy, or if she was in her early forties and thus faced an increased risk of having a baby with Down's syndrome (mongolism), it would be safer to use an egg from a screened donor. It would be like AID in reverse.

Genetic risks also could be reduced if a doctor examined the embryo before implanting it. Since it is possible for him to shave a few cells off an embryo without damaging it, he could analyze the embryo for sex and other genetic features before the pregnancy even began. Indeed, this has been used as a method of sex selection with rabbits and cattle.[30]

Embryo implants also could help women overcome such health barriers as uterine abnormalities, kidney ailments, and heart disorders. A woman who could not safely be pregnant could have her embryo transplanted into someone else, or have her egg removed, fertilized in vitro, and then implanted in a "host" mother.

It might even be possible someday to implant aborted embryos. Some women who suspect they are a few days pregnant have menstrual extractions, where the tiny embryos are removed by suction. Some of those embryos might be suitable for implanting in someone else. Perhaps, as in animal experiments, they could be stored frozen until we could find a new host.

Embryo banks could work like sperm banks. Before the husband or wife was sterilized, a couple could have their embryos stored for later implantation. (And even without embryo banks, a woman could have her oviducts tied and still be able to supply eggs for in-vitro fertilization and bear her children by embryo implantation.) Other bank customers might be young women who know they would soon be exposed to large doses of radiation, or who wanted children late in life but did not want the greater risks of such disorders as Down's syndrome. A further safeguard for all the customers would be to screen the sex and genetic health of the embryos that were stored.

Human embryo banks are far down the road, however, and they offer few advantages over sperm banks. For sperm banks already provide a reasonable assurance of safety and a realistic chance that the samples would later produce children. And they offer more flexibility than embryo banks because they directly involve only one party.

Artificial Wombs

If we look far into the future, imagining that our scientists have solved all the technical problems, we might envision a time when an artificial womb is the safest place for a baby to grow. No longer would an embryo or fetus be the helpless victim of the mother's German measles, uterine abnormalities, automobile accidents, drug misuse, malnutrition, kidney disorders, Rh incompatibility, heart disease, and the like. Floating serenely in their incubators of glass and steel, those fetuses would escape the risks and trauma of birth. And it would be easier for doctors to examine them and treat their disorders.

We can already see some of the benefits, for we are dramatically improving our methods for helping premature fetuses. For example, we are learning how to handle the most common cause of death in premature babies—respiratory distress syndrome, which results from underdeveloped lungs. One method, a perfusion system that reoxygenates the blood, has kept infants alive as long as ten days while giving their own lungs time to develop.[31] And another system, which ventilates the lungs, has greatly increased the chances for such infants not only to survive, but to be normal.[32]

Much further in the future is the possibility of using artificial wombs to change the way embryos and fetuses develop. It may be, for example, that fetuses maturing at slower rates would develop larger and more complex brains. Perhaps we could control this rate by diet and hormone treatments. But that would be futile as long as we have children in the usual way, for our evolutionary history has imposed an important limit: The fetus's head must be small enough to squeeze through the birth canal. An artificial womb (or routine cesarean sections) would remove that restriction.

And maybe there are other ways we could stimulate fetuses to develop their minds. It should be easier to teach fetuses in artificial wombs than inside uteri, but some of the problems would be the same. One writer has explained how his "cognofoetal friend Daedalus" is developing a system for natural pregnancies:

Sound heard underwater is distorted in a plummy, boomy way, but Daedalus is devising a pre-distorting circuit to deform the signal in just the right inverse way to make it sound correct in the womb. . . . He will try to prevent misuse of his technique (e.g. by advertisers who would happily pay mothers for the chance to teach their offspring to associate some product with the cosy security of the womb), and intends to integrate it with the education system. But he still cannot think of a good way of conducting inter-uterine examination.[33]

LEGAL ISSUES

As in-vitro fertilization and embryo implantation become more common, we will face some ticklish legal problems. For one thing, we will have to resolve who the parents of those babies are. If the husband and wife furnished the sperm and egg, and the wife bore the child, this would not be an issue. But what if the egg, or sperm, or both, came from donors? Here we ought to take the same approach as with AID: The husband and wife should be the legal parents if all parties consented to the procedure in advance and it was performed by an authorized person.

But who will we authorize to do these procedures, and in what situations?

While Steptoe and Edwards proceed with their experiments in England, and similar research is in progress elsewhere, researchers in the United States face a murky situation. A federal order in 1975 barred the Department of Health, Education and Welfare from funding any in-vitro fertilization research until it was evaluated by a national ethics advisory board. A panel of twenty experts recommended that "no research involving implantation of human ova which have been fertilized *in vitro* shall be approved until the safety of the technique has been demonstrated as far as possible in subhuman primates."[34] The ethics advisory board was not formed until 1978, and it remains to be seen what projects they will support. Jaroslav Marik, a Los Angeles obstetrician whose work was halted by the federal order, said he has a file of fifty women who want to try in-vitro fertilization and embryo implantation as their only hope to have children.[35]

Women who bear children for others will pose other legal headaches, some of which we discussed in chapter 2. Perhaps the only solution here will be to allow the parties to sign a contract in advance specifying, among other things, the financial arrangements, who will do the medical procedures, and who the parents will be. The parents then would assume full responsibility for the child, unless there were negligence by the host mother, or malpractice by the physician.

Regardless of the arrangement, the children should have the same legal rights as children conceived and born in the usual way. Indeed, we ought to replace the legal concept of "legitimate" children with the concept of "legal parents." For one thing, this might help reduce the pointless stigma attached to "illegitimate" children. In addition, this change would fit nicely with adoption procedures, and it would remove the unnecessary dilemma of whether parents should make false declarations when their children are conceived by AID or adultery.

While some of those legal problems loom on the horizon, other issues have already arrived. In 1978 Doris Del Zio and her husband filed a $1.5-million lawsuit against Columbia Presbyterian Medical Center and its chief of obstetrics, who terminated an experiment five years earlier in which her egg was to be fertilized in vitro by her husband's sperm, cultivated, and then implanted in her. (Incidentally, the physician who attempted the fertilization was Landrum Shettles, whose controversial methods for sex preselection we discussed earlier.) As compensation for her emotional pain and suffering, the jury awarded her fifty thousand dollars.[36]

Experiments with human embryos and fetuses are another legal minefield. In 1973 the United States Supreme Court implied that embryos and young fetuses do not have the full status of persons, ruling that a state may not protect fetuses from abortion until they reach the age of viability (about twenty-four to twenty-eight weeks).[37] And while some states prohibit experiments with live fetuses except to save them, the Supreme Court struck down a Missouri law requiring doctors to use as much care to save aborted fetuses as fetuses intended to be born alive.[38] So for now, experiments with eggs, sperm, embryos, and previable fetuses (younger than twenty-four weeks and weighing less than five hundred grams [about one pound]) are generally allowed if the other parties consent.

That, of course, does not settle the ethical issues. Let us consider them now.

ETHICAL ISSUES

The Status of Human Embryos and Fetuses

Our abortion laws imply that embryos and previable fetuses do not have all the usual human rights, but that does not necessarily mean they have no rights at all. Nor does it resolve the moral issues. For here we must ask: What rights *should* embryos and fetuses have?

Our answer depends heavily on what status we consider human embryos and fetuses to have. Perhaps we should first ask whether or not they are forms of human life. If we accept the usual criteria for "life" (cell reproduction, metabolism, growth) we find that embryos and fetuses qualify. And a geneticist would certify that they are human. So here we have a reasonably clear answer: Embryos and fetuses are forms of human life.

But that brings us to a more difficult, and crucial, question: Are embryos and fetuses human beings? Or, to put it another way, at what point in development does a human being emerge?

Before we can answer that, we have to decide what we mean by "human being." A dictionary definition is "a member of the genus *Homo*, and especially of the species *Homo sapiens*."[39] Now if we accept this definition, using biological classification as our sole criterion, we probably would conclude that human embryos and fetuses qualify as human beings. On the other hand, we might argue that this definition is too broad, for it could be stretched to fit a human "vegetable," or even someone whose brain is long deceased. If we narrowed the definition to mean a *functioning* human being, we would include what Salvador Luria has called the unique quality of man—the power of conscious thinking.[40] If we accepted this criterion, we would not regard an embryo, or any individual without significant brain activity, as a true human being. (What the minimum level of conscious thinking ability should be is arguable; Joseph Fletcher has proposed that an individual who has an IQ of 20 or less does not qualify as a person.[41])

Now we return to the basic question: When does a human being begin? Down through the centuries many different answers have been given, but they all have shortcomings. Let us briefly examine them.

Some people believe a fetus becomes a human being at birth. Here the baby leaves the cozy comfort and protection of the womb and must begin to function on his own. While this is a convenient criterion, we can see obvious deficiencies. One complication is that fetuses are born early, late, and on time. Do we say they all become human beings at the moment of birth, regardless of their level of development? And as we learn how to save fetuses at progressively earlier stages of development, birth will become a less critical event. Indeed, someday we may have artificial wombs, so fetuses won't be "born" at all. When would our "test-tube" fetuses become human beings? Would they be any less human beings than fetuses born in the usual way?

Others believe the fetus becomes a human being at the time of viability, when there is a chance of surviving outside the mother's body. This point currently is in the range twenty-four to twenty-eight weeks, but it will get shorter as research continues. So with this definition, our improving technology will mean that fetuses will become human beings faster in the future. And if a total artificial womb is perfected, "viability" will lose much of its meaning.

Another view, which dates back at least to the time of Aristotle, is that a human being begins at quickening, when the mother feels the fetus move. One problem here is that this depends on the mother as well as the fetus; quickening happens about the fourth or fifth month of pregnancy, but women pregnant for the first time may feel no movement, or misinterpret it, until after the sixth month. Another difficulty is in justifying muscular activity as the crucial criterion for defining a new human being.

We are coming to use brain-wave activity to determine when human beings cease to exist, so there is a certain logic in using the same criterion to define when human beings begin to exist. Moreover, this fits the idea that conscious thinking ability is a crucial quality in a human being. But this "solution" opens up another, equally bewildering problem: How do we decide when "brain life" begins? We do not know enough about fetal brain development to have a clear answer. Indeed, we might choose any time from about the fortieth day, when the basic structure of the brain is outlined and some brain-wave activity may be detected, until the twenty-eighth to thirty-second week, when the cerebral cortex, the part of the brain responsible for thinking, is well developed.[42]

Moving back on the time scale, we could consider a human being to emerge when the embryo implants in the uterus. With its genetic blueprint along and a suitable place to develop, the implanted embryo is all set for its long journey to babyhood. This criterion is convenient in that it eliminates ethical objections to birth control methods, such as IUDs and "morning after" drugs, that prevent implantation. But convenience is a poor basis for deciding when a human being begins. Another drawback is that research may eventually develop a substitute for implantation.

Finally, we might say a human being emerges when the egg is fertilized.

Here all the genetic instructions are in place for the egg to become a baby. As we will see in the next chapter, however, the origin of a human being could be pushed back even further, for there is a possibility, albeit a remote one, that eggs could develop into babies without being fertilized at all. Another complication with this definition is that body cells also carry all the genetic information to specify a baby.

In short, there is no clear answer as to when a human being begins to exist. That is obvious enough, for people have been arguing about it ever since the fall of man. If we consider "a human being" synonymous with "a form of human life," we might well choose the fertilized egg as the starting point. But if we think we have to think in order to qualify as human beings, some level of brain activity seems the logical criterion.

Experiments with Embryos and Fetuses

As in-vitro fertilization and implantation experiments with human embryos become more common, we need to face several ethical issues. One problem is that some embryos are discarded when the experiments are over. Is that immoral? It largely depends on whether we consider embryos to be human beings; if they are not, they do not necessarily have the right to life. On the other hand, those who consider embryos as human beings, or at least believe they should have the right to life, face the difficult task of where to draw the line (if anywhere) on abnormal embryos having a full right to life.

Another issue is that embryos cannot consent to be the subject of experiments. Here again, whether they should have that right depends largely on the status we consider embryos to have. Yet we also must recognize that this particular problem is not unique to "test-tube babies." Indeed, it is hardly stretching the point to say that many conventional pregnancies are also experiments; the fetus may be endangered by the mother's health, nutrition, use of drugs, physical activity, and genetic condition. In fact, about 5 percent of all babies are born with genetic disorders—over one hundred thousand children each year in the United States alone. Yet genetic counselors report that some parents will take a chance even if they know there is a 25-percent risk of having a child with a serious abnormality. Embryos and fetuses have nothing to say about those experiments, either.

Abnormalities and laboratory errors are another concern, for we might expect that manipulating eggs, sperm, and embryos in the laboratory could cause some damage. In-vitro fertilization, however, seems fairly safe. Eggs before and after fertilization in the laboratory look normal under the microscope.[45] Still, the clearest evidence of safety would come from examining the offspring. Reports of animal experiments do not indicate an unusual rate of abnormalities (except for one study with rats where there was a high incidence of unusually small eyes[44]). But the data are scanty. Indeed, there have been no nonhuman primates produced from in-vitro fertilization.

And with humans, there are only two babies to examine so far. They both look normal.

Are the risks small enough to proceed with experiments involving humans? At the Del Zio trial, a New York gynecologist testified that the risks were "no greater than in normal pregnancy, in fact less . . . because we could examine (the embryo) microscopically before putting it into the uterus" to make sure it was normal.[45] But not everyone agrees. Researchers usually start with simple organisms and gradually work up to monkeys and apes before trying new procedures on people, but this was not done with in-vitro fertilization and embryo implantation. Indeed, Leon Kass, a biochemist and ethicist now at the University of Chicago, proposed a moratorium on such experiments with human embryos, at least until the risks are thoroughly assessed in animal studies, including monkeys.[46] And Paul Ramsey has raised another, even more basic, objection: "The decision must be that we cannot rightfully *get to know* how to do this without conducting unethical experiments upon the unborn who must be the mishaps (the dead and the retarded ones) through whom we learn how."[47]

How severe are those risks? Although the skimpy evidence we have indicates these methods are moderately safe, there is no question we need more extensive animal studies. Even those experiments, however, cannot guarantee safety with humans, as Ramsey has pointed out. Yet we can never have that assurance in advance, for there is some uncertainty whenever treatments are first tried on people. Unless we can rely on professional judgments, supported by reasonable criteria, to determine when new methods are safe and effective enough to try with humans, we can kiss further medical progress good-bye.

In 1966 Robert Edwards wrote: "If rabbit and pig eggs can be fertilized after maturation in culture, presumably human eggs grown in culture could also be fertilized, although obviously it would not be permissible to implant them in a human recipient.[48] Now Edwards and Steptoe, and others outside the United States, are doing those experiments that "obviously" were not "permissible" in 1966. And as this research continues, we will increasingly accept these methods if they prove reasonably safe. Indeed, a survey of eighty-eight unmarried women showed 66 percent would be ready to use in-vitro fertilization, with their eggs and sperm from their husbands, if this were their only way to become pregnant; with donor sperm, the figure dropped to 11 percent.[49]

Experiments to grow babies entirely in the laboratory, however, are a different matter altogether. First and foremost, the technology for doing this is far more complex, and much less developed than for in-vitro fertilization and embryo implantation. And even if we knew much more than we do now, an embryo would spend so much time in an artificial womb that its chances of being damaged by a laboratory error would be high. Moreover, about one-fourth of all natural conceptions are spontaneously aborted, at least a third of them abnormal; this natural screening mechanism for healthy babies would be absent in an artificial womb.

So here the verdict is clear: because of the great risk of abnormality, and the respect that human embryos and fetuses deserve (regardless of whether we consider them human beings), any attempts to grow human babies from embryos in the laboratory would be unethical. Before there could be any acceptable standard of safety, we would need far more animal testing, including research with nonhuman primates.

But there is another dimension to artificial womb experiments: Many aborted and premature fetuses with no hope of survival are available for research, and such work could help save other fetuses in the future. That is no idle claim. While most fetal research is done on the tissues of dead fetuses, research with live fetuses has been instrumental in developing German measles and Rh vaccines, and better methods to help infants with breathing problems. These advances already have saved tens of thousands of lives.

Nevertheless, the idea of doing research with live fetuses has stirred up considerable controversy. In 1975 the Department of Health, Education and Welfare adopted rules that apply to all projects funded by that agency; those rules permit research on unborn or newly aborted fetuses but ban studies that would end the life of a fetus or would keep a fetus alive artificially to prolong an experiment. It is reasonable to allow experiments within these rules that are substantive, well designed, and must be done on live fetuses, but it would clearly be unethical for researchers to participate in any abortion decisions that would provide fetuses for their experiments. Indeed, the general practice of abortion, especially for reasons of convenience, poses more severe ethical problems than does the use of aborted embryos and previable fetuses in research.

As our technology advances, however, we will increasingly face another dilemma: Which embryos and fetuses should we try to save? One side of the question is how to deal with abnormalities, for someday we will not have to accept the natural screening action of the womb; we will be able to save very premature fetuses, a disproportionate number of which are abnormal. In fact, we are already encountering this problem. Several hospitals have reported that premature babies weighing less than 3.3 pounds that survive have at least a 50-percent chance of being handicapped.[50] We will learn how to prevent some of those disabilities, and some we cannot prevent will be minor, but that does not erase the problem of deciding where (and whether) to draw the line.

And when our technology brings us artificial wombs, enabling us to save aborted embryos and fetuses that are otherwise normal, should we do it? Why not? Could we justify a negative answer because we needed to reduce our population growth, or because their biological parents did not want them? I doubt it. A more likely limit would be coldly practical: We would have to decide how much of our finite medical resources we could spend on trying to save aborted embryos and fetuses. Or perhaps the justification would be one that is used for abortions: Embroys and young fetuses are not human beings, so they do not necessarily have the right to life. Yet in other ways the two situations are not the same, for aborted embryos and fetuses do not compromise the mother's right to privacy, nor threaten her physical or mental health.

Many people who oppose abortion on demand also oppose fetal research. That is understandable, but it is also ironic. Such research will have the long-term effect of saving fetuses that otherwise would die, and that could make abortion decisions even more difficult in the future. For someday, a decision to abort may require another, separate decision on whether to help the aborted embryo or fetus become a full-fledged human being.

SOME IMPLICATIONS

As basic research tools, in-vitro fertilization and embryo implantation will bring us new insights that could lead to better contraceptive methods, new solutions for infertility, and fewer abnormal babies. And this method of producing children, if it proves reasonably safe and effective, offers compelling benefits to couples who are infertile, or who carry genetic disorders. At the same time, however, this technology opens the door to other applications that are not as clearly desirable.

One concern is the prospect of women bearing children for others. There could be many reasons—physical, psychological, career plans, vacation plans—for hiring a host mother. But those reasons are not equally good. It is one thing for a woman who could not safely bear a child to seek a host mother; it is quite another matter if she could not bear the inconvenience. But where do we draw the line? Or should we even try? Once the technology is available, we will find it difficult to prevent applications we dislike, for there will be honest differences of opinion about what is appropriate. As with abortion, the law probably will allow each person to follow his own moral drummer.

In 1970 a British embryologist suggested a fee of £2,000 per pregnancy for host mothers.[51] Adjusted for the inflation in Britain, the exchange rate of the pound, plus 11 percent for fringe benefits (including paid-up health insurance and workperson's compensation), the current price works out to $15,618.81, plus tax and tip. That is expensive enough that only wealthy people would be likely to use these services—unless we could talk Blue Cross into covering it.

Some people have charged that this would be a way for the rich to exploit poor women. But the host mothers could form a labor union to ensure that their prices and working conditions were satisfactory. Besides, there are women who might genuinely enjoy offering their wombs for rent. According to one report: "a few women . . . have called our office (though we are not engaged in work of this nature) to enquire whether they might volunteer their services should such ventures become reality. They state that they love being pregnant, and would arrange to always be in this condition if it were not for the matter of having to keep the babies. They think that hiring out their uteri would be a fine way to make a living."[52]

One concern with surrogate mothers is that they could weaken family ties. The eventual mother, and the rest of her family, would miss the experience of having their baby become part of their lives even before birth. And when the physician cut the umbilical cord, he also would sever the relationship between the newborn and his host mother, leaving the baby to start over, forming new bonds with his legal parents. Indeed, a curtain of anonymity might have to separate the first and second parents in order to make things easier —especially for the adults.

And if we ever have total artificial wombs, the fetus will not necessarily develop a bond to anyone. But science will come to our rescue, for their needs could be handled mechnaically. In one study, for example, premature babies developed faster, both physically and mentally, when they were placed in a "mechanical mother," complete with rocking motion and the sound of a beating heart.[53]

But there's a problem here: Relationships are not objects. Without human relationships there is a void that cannot be filled, even by the most sophisticated simulation techniques. Indeed, the very idea of replacing the human touch with laboratory treatments brings us to the chilling specter of *Brave New World.* And it brings us to the greatest danger of these laboratory methods: They could depersonalize reproduction to the point that someday we would regard people as manufactured objects. That would be tragic—for all of us. As one writer remarked:

Once a woman has no more difficult or lengthy role in reproduction than a man . . . she will find that society does not expect her to have a special relation to her offspring that takes up years of her life, and also she will not expect it of herself. Too, a society that can grow fetuses in a laboratory will be more disposed to have meaningful day- and night-care centers and communal nurseries on a large scale, for the state, being a third parent, will wish to provide for the maintenance and upbringing of its children.[54]

Artificial wombs could free all women from the experiences of pregnancy and childbirth. Although we might view this prospect with distaste, we should not underestimate how fast social attitudes can change. Not too many decades ago people vehemently opposed the use of anesthetics during child-birth; they were "unnatural." Now we accept that there is a proper place for anesthetics, drugs that induce labor, cesarean sections, fetal monitors, and other intrusions on nature. So it may come to be with artificial wombs. Indeed, someday our descendants may regard pregnancy and childbirth as a risky and crude way to have children.

But total artificial wombs are far in the future. And even when all the techniques are worked out, there will still be a major barrier to their routine use—cost. Natural mothers will be less expensive. So it is unlikely that a society would use this technology except for special situations—unless some government made it a high priority for funding.

Could that ever happen? After Daniele Petrucci announced that he had grown human embryos in the laboratory for several weeks, the following editorial comment appeared in the Chinese newspaper *Jenmin Jin Tao*: "These are achievements of extreme importance, which have opened up bright perspectives for similar research. . . . Nine months of pregnancy is no light or easy burden and such diseases as poisoning due to pregnancy are detrimental to health. If children can be had without being borne, working mothers need not be affected by childbirth. This is happy news for women."[55]

One egg, one embryo, one adult—normality. But a bokanovskified egg will bud, will proliferate, will divide. From eight to ninety-six buds, and every bud will grow into a perfectly formed embryo, and every embryo into a full-sized adult. Making ninety-six human beings grow where only one grew before. Progress.

Brave New World by Aldous Huxley

Xeroxing of people? It shouldn't be done in the labs, even once, with humans.

Robert Francoeur

4

Cloning and Virgin Birth

First we learned how to have sex without having babies; now we are learning how to have babies without having sex. But with artificial insemination, test-tube fertilizations, embryo implants, and artificial wombs, one thing remains the same: It still takes a woman and a man to make a baby.

Even that could change. We know of ways to make babies that have just one genetic parent—male or female—and we might be able to make as many genetic copies as we like. This opens the door to such delights as identical centuplets, a unisex society (more likely female), and mass-producing people with special features our brave new society wanted.

Let us begin with the state of the art.

TECHNOLOGY FOR MULTIPLE COPIES

Parthenogenesis

Parthenogenesis (literally, "virgin birth") is a process where an unfertilized egg starts to divide and develop as an embryo. If all goes well, the end product is a baby whatever. It happens with sea urchins, frogs, turkeys, chickens, and many insects. For some, especially certain insects, it is the usual way to reproduce; in others, such as chickens and turkeys, natural parthenogenesis is rare.

Now we are learning how to cause artificial parthenogenesis. With frog eggs, all it takes is a pinprick to start embryonic development. Mammalian eggs need shock (electric, heat, cold) or chemical treatments. In 1939 Gregory

Pincus, who later directed the Worcester Foundation for Experimental Biology and helped develop the birth control pill, reported that he had induced parthenogenesis in rabbit eggs, and a few of them had developed into baby rabbits.[1] But some scientists are skeptical, for despite many other attempts, no one else has been able to send out that kind of birth announcement. Apart from Pincus's work, the survival record for mammals is shared by two mouse embryos that developed normally for just over half the pregnancy period.[2]

Many people, of course, believe a virgin birth has graced the human race. Perhaps it has. But if parthenogenesis is the way it happened, some of our historians must be male chauvinists. For in humans, this method would produce only females.

Parthenogenesis produces babies with 50 percent, 100 percent, or some other proportion of chromosomes compared with their genetic parent. Unfertilized eggs have only half as many chromosomes as body cells, so the first figure makes sense. The second figure usually means there is a double dose of one set of chromosomes, though it is also possible to have the same two sets of chromosomes as the mother. When only one set is present, however, in a single or double dose, the baby resembles the mother but is not a genetic duplicate. The other figures, ranging from less than 50 percent to well over 100 percent, reflect various chromosome games the egg can play. But whatever the game, no uninvited chromosomes can crash the gate. And every night is ladies' night.

Although human parthenogenesis seems a long shot, betting against science is a way to get rich slow. Indeed, Jean Rostand, an eminent French biologist and writer, predicted that parthenogenesis "will someday be possible, so that we will be able to have as many exact copies of an exceptional individual as we want."[3]

But don't hold your breath on it.

Cloning

Cloning is a way to grow many identical cells or organisms from a single ancestor. In the 1950s F. C. Steward of Cornell University showed how to do it with certain plants. By separating the cells of a carrot's root and suspending them in a growth medium, he produced new carrots by the thousands.[4]

Copying animals is more difficult, but sometimes nature does it herself. For example, we produce identical twins or triplets when a young embryo splits into two or three embryos, each of which develops into a baby. Some animals do this routinely, and armadillos are the prime example among mammals. The nine-banded armadillo regularly has identical quadruplets, and the mulita armadillo usually produces eight or nine, and as many as twelve, identical offspring.[5]

How they do it is a mystery, but the answer might give us the same capability. It might be a bit hard, though, to find volunteers to bear a dozen or so babies at a time. So we would need to handle some of the steps in vitro. We could let an embryo develop a few days in the laboratory, separate the cells, and grow each one into a baby in artificial or natural incubators. Perhaps we could even take each separated cell, let it develop awhile, separate each of those cells, and repeat the cycle as often as we wanted. And we could freeze the cells until we were ready to run off another batch of people.

Could it happen? Scientists have separated cells of sea urchin embryos at the two- and four-cell stages and developed them into whole embryos.[6] They've also done this with two-cell embryos from frogs and rabbits. The current record is to separate the cells from a four-cell mouse embryo, develop each in vitro, implant them, and produce three mice.[7] It is also possible to take one cell from a four-cell embryo, let it divide once, separate those two cells, and develop them into mice.

In *Brave New World*, which was written in 1932, Aldous Huxley imagined that we could let an embryo reach a ninety-six-cell stage, then separate the cells and grow each one into a person. But he also implied there is a limit.

And indeed there is. From a single fertilized egg must come the specialized cells for our liver, kidneys, heart, skin, brain, blood, and all the rest. Although each type (except germ cells) has the same genetic information, individual cells use only a portion of their total information, and they use different pieces. After an embryo develops awhile, cells become locked into specific patterns of using that information; as they become more specialized, cells lose their ability to produce the wide variety of cells that make up a whole organism (or at least we do not know how to restore that ability). If we did, almost any body cell could produce a baby.

If the cells are not too specialized, however, there is a way to turn the clock backward and then forward: The cytoplasm of an egg can coax a nucleus into setting its genes free to express themselves in myriad patterns. This type of cloning amounts to a nuclear transplant.

The idea is simple: Once the nucleus of an unfertilized egg is destroyed or removed, the egg is ready to obey a nucleus from a body cell, which has the same number of chromosomes as a fertilized egg. The renucleated egg becomes like a fertilized egg, except that all its genetic information comes from a single animal. The nucleus of an unfertilized egg may be destroyed by radiation, or removed either by a tiny probe or by treatment with a substance (cytochalasin B) produced by a fungus. The egg may then receive its new master by microinjection, or by fusion with a body cell. If all goes will, the egg will supply the nutrients and machinery to convert the genetic instructions into a baby with the same genetic features as the donor of the nucleus, including sex. Since a donor could supply almost any number of nuclei, we could make as many copies as we wanted.

It is not just a theory. This method of cloning has produced frogs, newts, and fruit flies. And the list will grow. John Gurdon, now at Cambridge University, and his coworkers have done the most extensive work. They used ultraviolet radiation to destroy the nuclei of unfertilized frog eggs. Then they used a tiny pipette to remove nuclei from the intestinal cells of tadpoles and transfer them into the irradiated eggs. A few of those eggs developed into mature frogs. Nuclei from adult frogs generally did not work. The few that did (from skin cells) produced tadpoles that failed to reach adulthood.[8]

Although cloning frogs is difficult, it is much simpler than cloning a human. Human eggs are not readily available, and there is little reason for a woman to volunteer for surgery when the researcher wanted to do an experiment. And in published work, no one has yet shown that it is possible to clone a mammal by using a body cell nucleus from an adult. Another problem would be the nuclear transfer. Human eggs are a thousand times smaller than frog eggs, so they would be more vulnerable to damage from a pipette. Although this may be solved by fusing the egg with a body cell, similar experiments with mouse eggs have not produced eggs that develop beyond the two-cell stage.[9] The most successful experiment with mammals so far, involving the transfer of nuclei from rabbit embryonic cells into unfertilized rabbit eggs by fusion or microsurgery, produced a few apparently normal rabbit embryos that reached at least the eighteen to twenty-six-cell stage.[10] There was no attempt, however, to implant them. Yet even if a nuclear transfer were done successfully on a human egg, the embryo would need a place to develop into a baby—either an artificial womb or a woman willing to supply her own. The technology for a complete artificial womb is not yet in sight. And implanting human embryos was tried for a decade without success (the claim of Douglas Bevis notwithstanding) before Steptoe and Edwards made their announcement in 1978. But despite the birth of one "test-tube baby," the odds for success in each implantation attempt remain small.

Oh yes, there's one other report. In 1978 David Rorvik, a science journalist, wrote a book alleging that a sixty-seven-year-old millionaire bachelor had himself cloned. Rorvik said he helped arrange the experiment, in which the nucleus of each of several unfertilized eggs was removed and replaced, by means of fusion, with a nucleus from one of the millionaire's cells; one of the eggs then developed in the laboratory for about five days and was implanted in a seventeen-year-old virgin, who gave birth to a baby boy in December 1976. As with Bevis, Rorvik offered few details of the experiment and refused to identify the participants, saying he was obliged to preserve their privacy. His book sets new standards for the label "nonfiction."[11]

A different version of cloning has already been developed for a mammal. Clement Markert of Yale University devised a way to remove one set of chromosomes—either those from the sperm or those from the egg—just after fertilization. By chemical manipulations he can coax the remaining set to double without the egg dividing, producing an egg with a double set of the egg's (or sperm's) chromosomes. Such an egg has the same number of chromosomes as a fertilized egg and can begin embryonic development.[12] Using that method of microsurgery, developing the embryos a few days in the laboratory, and then implanting them in female mice, Peter Hoppe and Karl Illmensee at the Jackson Laboratory in Bar Harbor, Maine, produced seven baby mice, all females. (This method produces only females, even if the sperm's genetic information is used, because doubling one set of chromosomes cannot produce the XY combination necessary for males, and YY combinations are not viable.) Two of those mice received all their genes from the sperm; the others inherited their genes solely from the egg.[13]

None of the seven mice were clones of their genetic parent, but if the same procedure were repeated on those seven mice (retaining the chromosomes of their eggs), their offspring *would* be clones. With this method, according to Markert, the cloning of mammals "should be possible very soon."[14]

What about cloning humans? The technical problems are daunting, but scientists will solve them eventually. James Watson, Nobel laureate for his work on the structure of DNA, predicted in 1971: ". . . if the matter proceeds in its current nondirected fashion, a human being born of clonal reproduction most likely will appear on the earth within the next twenty to fifty years, and even sooner, if some nation should actively promote the venture."[15]

IMPLICATIONS

Should some nation "actively promote the venture"? Let us survey the possible benefits from parthenogenesis and cloning, assuming that the technical problems will be solved.

First of all, these techniques are valuable research tools. They can help us probe such mysteries as how embryos develop specialized cells; whether the egg or sperm carries viruslike substances, some of which may cause cancer; where specific genes are located on chromosomes; and the way certain chemicals cause mutations.[16] Cloning animals may help scientists locate disease genes. And multiplying cells by cloning could be a way to grow material for transplantation, or for medical diagnoses. (We will see examples in the next chapter.)

As far as using these methods to produce babies is concerned, one use would be to provide children to infertile couples, or to unmarried people.

With both parthenogenesis and cloning, the parent(s) would know in advance the sex of the child. And if these methods were safe, people could use them to avoid high genetic risks. If a husband or wife had, or carried, a genetic defect, the spouse could be cloned. If they only carried the same disorder but did not display its symptoms themselves, cloning either one would eliminate their risk of having a baby with that defect. Their child would be a carrier, like them.

Parthenogenesis and cloning eliminate the requirement for men. Women could have their eggs treated and implanted, always producing girls. Women's liberation would be complete. On the other hand, if we developed artificial wombs and learned how to culture cells to provide an endless source of unfertilized eggs, cloning would make either sex optional.

Another use, especially of cloning, is to grow many genetic copies of an individual. In animal breeding, it would be a way to multiply a prized animal. While E. S. E. Hafez of Washington State University has shown how it is possible to make cows produce multiple offspring from a single embryo,[17] cloning by microsurgery would be a more efficient way to mass-produce animals with fixed genetic features. According to Markert: "It takes decades, even centuries, to produce a particular kind of dairy cow. We could produce one in a year or two [with cloning]."[18]

We also could make large batches of human clones. One use would be to provide control groups for statistical studies. Perhaps our geneticists, psychologists, and sociologists could settle, for once and for all, the relative importance of heredity and environment in human intelligence and behavior.

We would probably choose to replicate the healthy, the beautiful, and the bright (assuming, of course, that you and I were willing to donate some of our cells). Perhaps we would also like a few copies of Billie Holiday, or Michelangelo, or Albert Einstein, or Mahatma Gandhi. It is too late to replicate them, but we could start freezing cells from our current heroes. Then we could take as much time as we wanted in order to decide whether they were truly valuable enough to copy.

Our shopping list would not have to stop there. We also could copy dimwits to do dull, repetitive tasks. And we could replicate people especially endowed to be superwarriors, deep sea divers, olympic athletes, or ditch diggers. The possibilities are limited only by our imagination.

Cloning could provide a sort of immortality. People might have themselves copied at regular intervals to ensure that they will be with us forever, in body if not in soul. Reigning monarchs need not worry about a successor to the throne. Astronauts could embark on long voyages, and their duplicates would be on hand for the landing. In another version of immortality, a person could have his clones frozen at the ripe young age of sixteen; they would be his source of rejection-proof body parts.

The applications can get as far-fetched as we like. But now let us return to reality.

SOME CONCERNS

Safety
We might expect that nuclear transplants and shock or chemical treatments could damage the eggs. They do. In the cloning experiments with frogs, over 90 percent of the eggs with new nuclei did not divide at all, or developed abnormally. Furthermore, nuclei from adult body cells have not yet produced even one adult frog. And with the microsurgical method Hoppe and Illmensee used, it took hundreds of treated eggs to produce a total of seven baby mice.

Although scientists are refining the techniques, the verdict for now is clear: It would be improper to try to produce humans by cloning, at least until the methods prove reasonably safe and effective in animals, including primates. The current risks of abnormality and our reverence for human life should rule those experiments out.

The conclusion is the same with parthenogenesis, for abnormal development is the rule with mammals. Not only do activated eggs fail to develop into live offspring, but microscopic examination of such eggs shows many abnormalities.[19]

Another danger with human parthenogenesis (and cloning by Markert's method) is the lack of genetic variety in the embryos. Each of us carries an average of three to eight defective genes, but most are recessive (which means we will not have the disorder if a defective gene is paired with a normal one). Since we have a hundred thousand or so different genes, the odds are small that we would happen to receive the same defective gene from each parent, though that risk escalates when parents are blood relatives. But in parthenogenesis there is no chance for the genes from one parent to compensate for defective genes from the other; all the recessive defects in an egg with a single, double, triple, or some other dose of a single set of genes will show up in the offspring.

We could avoid that problem in animal experiments by using eggs without any defective genes. When we inbreed animals for many generations, discarding those that produce defective babies, the survivors can inbreed safely and are not likely to carry severe recessive defects. Even better than their eggs, however, would be the eggs from healthy animals produced by Markert's microsurgical method (which is used on highly inbred strains).

In humans, though, these methods have not attracted a wave, swarm, bevy, brood, flock, herd, pack, drove, gaggle, or even a covey of volunteers.

The best egg donors would be women from a small group whose members have intermarried for many generations. But there is little reason for them to volunteer their eggs, for their chances of producing a healthy baby by this method are approximately zero. And deliberately inbreeding people for twenty generations or so is out of the question, for the participants might feel a touch of resentment at the restrictions on their freedom, and their high risks of having defective babies. Besides, six hundred years is a pretty long time to get ready for an experiment.

We could not ethically try to produce a human being by parthenogenesis, but we cannot stop nature's experiments. Some women have ovarian tumors containing bits of bone, teeth, hair, skin, fat, nerve tissue, connective tissue, and other specialized materials that apparently result from spontaneous parthenogenesis.[20] Those unfertilized eggs start developing into embryos, forming specialized types of cells, and then go no further (except that they may become malignant).

So human parthenogenesis may be more common than we had realized. No one knows for sure whether an unfertilized egg has ever developed into a person. Indeed, more than two decades ago there was a study of nineteen women who thought they might have had a daughter by parthenogenesis; only eighteen of those claims were definitely disproved.[21]

Nevertheless, a virgin birth by this method still rates as a miracle.

Multiple Copies

The old-fashioned way of reproducing has distinct advantages, besides the pleasures of trying. When a man and woman together produce a child, they forge a special bond with their child, and with each other, that helps provide a loving and stable environment for the family members to develop as persons.

Multiple copies, however, could strain those relationships. If a child were a genetic copy of someone his parents admired, he might feel a bit detached from them and attached to his genetic parent. And if the child were copied from one of the parents, he might feel less close to the other. He also might experience a novel type of sibling rivalry with his twin parent. According to Paul Ramsey: "Our children begin with a unique genetic independence of us, analogous to the personal independence that sooner or later we must grant them or have wrested from us. For us to choose to replicate ourselves in them, to make ourselves the foreknowers and creators of every one of their specific genetic predispositions, might well prove to be a psychologically and personally unendurable undertaking."[22]

Another, overwhelming problem is deciding whom to copy, and how many copies to make. It is not comforting to think of entrusting anyone with that power, but some person or group would have to make the decisions and select the criteria. For if it were left on a voluntary, individual basis, those copied would be mostly the wealthy, the powerful, and the egotistical. We have enough politicians already.

Intelligence is one criterion we would likely choose, but even that would bring problems. There probably are limits on how many brilliant people we could tolerate at any one time, for they would spew out a torrent of new and conflicting ideas that could dramatically alter our society. Yet for many people, our world already changes too fast.

Then there would be the disappointments. A genetic copy of Winston Churchill would not necessarily resemble the original in ambition, insight, leadership, courage, and other qualities. Although he might inherit certain tendencies and limits, his personal qualities would be molded by his environment. To make the system truly effective, we would have to copy all of Winston's colleagues, and their colleagues, and their colleagues, and recreate the crucial situations for each batch of copies, so that in the end Winston II might develop some of the same qualities as his older twin.

But if we reached the point of conditioning copies to have specific personalities, we might find it easier if they were a bit stupid, so they would not question what was happening to them. We could manufacture groups of dull people, each dressed alike and looking alike, who enjoyed going to work and doing the same, boring tasks day after day. We could condition our drones so they would not know what they were missing. They would be happy.

To be sure, we would reject that Brave New World, insisting that our genetic copies—like identical twins—would be free to develop their own personalities and interests. Still, making copies deliberately is not quite the same thing. Our reasons for doing so, whatever they were, would impose expectations on the new persons. And despite our noble intentions, we might feel a twinge of regret when Abraham Lincoln LXVII decided to open his Lincolnburger fast-food restaurant.

The greatest danger, in the long run, is that xeroxing people would turn them into objects. We might start thinking of them in terms of the $4.02 or so that their body chemicals are worth,[23] or measure their value in terms of how many goods they can produce versus how much it costs to produce them. In a world of regimented marionettes we could lose our notion that each person is unique and precious and is entitled to make his own free way in the world.

Genetic Diversity

Sex is a way for nature to try out new genetic combinations. Since individual eggs and sperm randomly carry one of the genes from each pair present in the body cells, no two sperm or egg cells, even from the same person, have the same set of genes. So nature rolls her genetic dice for each pregnancy, virtually guaranteeing that children from different eggs and sperm will be genetically unique. Copying individuals would change all that.

Throughout our long history, genetic diversity has helped us survive. Nature has counted among her victims many organisms that could not adapt to their surroundings. But species with a wide variety of genetic combinations have a special advantage: They are more likely to have individual members that can survive and reproduce when their environment changes.

It is something like a professional football team. A team would have a hard time indeed if all its members had the same size and shape (even seven feet tall and 260 pounds), for different positions require different skills. Teams with a variety of players and special skills would have more flexibility to "survive" an assortment of other teams and playing conditions.

We also can see the benefits of diversity in agriculture. Highly inbred strains may give better crop yields under certain conditions, but they are more vulnerable to surprises; diseases, insect invasions, or new weather conditions can devastate them. So we may produce more food in the end by using plants with lower potential yield but greater diversity.

Now if most of us were made from the same genetic blueprint, the human race would be more vulnerable to a sudden change in our environment. Suppose, for example, a new microbe evolved that resisted our medicines and killed a certain genetic type of person—and everyone was of that type. Our only hope would be our wits; perhaps we could figure out how to manipulate our environment to eliminate the problem. But even here there is a catch: Our intelligence is itself a product of genetic diversity. According to Bentley Glass, a distinguished geneticist and former president of the American Association for the Advancement of Science: "The intelligence of man is an evolutionary product of natural selection for adaptability to great variation of surroundings, to tremendous vicissitudes of experience. . . . [As] long as we have freedom to choose, so long will human intelligence based upon genetic diversity remain a primary ingredient."[24] So if we relied on genetic duplicates to be our intellectual leaders, entrusting our survival to that intelligence and voluntarily discarding our insurance policy—genetic diversity—we would discover (1) we had overestimated our intelligence because (2) along with our safety net we had thrown out our wire to walk on.

AN OVERVIEW

Cloning and parthenogenesis are valuable research tools to unveil the secrets of the cell. And cloning cells is helpful in medical diagnosis and treatment. Because many types of plants can be grown from individual body cells, we will also use cloning to copy the healthiest, fastest-growing, most disease- and insect-resistant plants. Already, it is profitable to clone expensive orchids; in fact, one method is patented. And although it is much more complicated to clone animals, we will probably find it worthwhile to clone certain valuable livestock as a way of expanding our food supply.

As methods for reproducing people, however, cloning and parthenogenesis are solutions in search of a problem. For if we scrutinize the possible benefits, we discover there is less here than meets the eye. Treating infertility, reducing genetic risks, sex selection, and providing children to unmarried people can all be handled by other methods that are better developed and safer. Replicating people we admired would be a fiasco, for their copies would have unpredictable personalities unless we could condition them thoroughly. And using people as a source of spare body parts, or as guinea pigs for statistical studies, does not deserve comment.

Brave New World is not just around the corner. In fact, the impact of this technology on our society will be less drastic than we might imagine. Our need for genetic diversity would limit how many people we would clone. And even if we wanted to copy only a few people, we would find our belief in the uniqueness, dignity, worth, and freedom of each person too high a price to pay. In *The Second Genesis: The Coming Control of Life*, Albert Rosenfeld has given us a glimpse of what it could be like:

In our current circumstances, the absence of a loved one saddens us, and death brings terrible grief. Think how easily the tears could be wiped away if there were no single "loved one" to miss that much—or if that loved one were readily replaceable by any of several others.

And yet—if you . . . did not miss anyone very much, neither would anyone miss *you* very much. Your absence would cause little sadness, your death little grief. You too would be readily replaceable.[25]

Although it is a human right for people to have their infirmities cared for by every means that society can muster, they do not have the right knowingly to pass on to posterity such a load of infirmities of genetic or partly genetic origin as to cause an increase in the burden already being carried by the population.

H. J. Muller

So God created man in his own image, in the image of God created he him; male and female created he them. And God blessed them, and God said to them, Be fruitful, and multiply, and replenish the earth, and subdue it. . . . And God saw everything that he had made, and, behold, it was very good.

Genesis 1: 27-28, 31

5

Genetic Dice

Nature uses an ingenious device—sex—to operate her genetic lottery, a giant wheel of fortune that spins out new genetic combinations into the world, both winners and losers. Each of us is a random combination of our parents' genes, and, like it or not, we have to live with the boundaries they set for us. Nature's game of chance is a hard fact of life, especially for the unlucky ones, and all we can do is play the numbers we get as best we can. It is a price we pay for genetic diversity.

The cost of diversity—genetic misfits—has always been with us, but gradually we have changed our way of paying the bill. Natural selection used to operate with ruthless efficiency, eliminating people with serious defects before they could reproduce. So for hundreds of thousands of years, we inched our way up the genetic staircase, keeping our best features and discarding the bad ones. And our physical evolution supplied the fuel for our cultural advancement. We discovered, for example, the advantages of living in groups and caring for the needs of each other. Indeed, we have made truly spectacular progress in saving our brothers and sisters from an early death. In doing so, however, we reduced the premium on genetic fitness.

This poses a problem. While natural selection now operates with a muzzle, we have accelerated our exposure to agents that can damage our genes (for example, radiation and certain drugs, dyes, food additives, and plastics). Joshua Lederberg, a Nobel laureate at Stanford University, has estimated that 80 percent of our mutation rate comes from environmental factors we can control.[1] So now we face a dangerous combination: more people are

49

acquiring defective genes while modern medicine is helping more of them survive and pass their genes on to their children.

Genetic diseases are a growing problem, as we can see from some statistics. More than two thousand diseases are at least partly genetic.[2] They afflict an estimated 150 million people and account for one-fourth to one-third of all hospitalizations of children.[3] According to some estimates, the number of children born with genetic deficiencies will double in the next five to ten generations.[4] Better medical care already has caused a three-to-four-fold increase in the number of surviving infants with Down's syndrome, a disorder that accounts for about one-sixth of all retardates.[5]

And these figures reveal only the tip of the iceberg, for they do not include people who carry recessive diseases. A common guesstimate (probably too low) is that, on the average, we each have three to eight defective genes, and each of us is almost certain to carry at least one.[6] For each person with a recessive defect, there may be as many as a hundred people who carry that disorder. So we have to multiply the number of people with genetic diseases by a large factor to estimate the total number of defective genes in the human gene pool. Not only is that number increasing, but so is the proportion. Indeed, our race harbors a genetic time bomb, slowly ticking away.

And these are just statistics and predictions. They cannot show the reality of a couple exhausting their emotional and physical resources on their hopelessly retarded and deformed baby. Nor can numbers represent the individual trying to cope with his disability. Some do beautifully, thanks to modern medicine, a loving environment, and their personal qualities. But others lead lives stamped with frustration, meaninglessness, and a sense of being a burden to those around them who care (and sometimes do not care) for them.

We face a genetic predicament: We want to improve the quality of individual lives, but we do not want to ticket increasing numbers of our descendants to genetic oblivion. Theodosius Dobzhansky, an eminent geneticist, summarized the dilemma: "If we enable the weak and the deformed to live and to propagate their kind, we face the prospect of a genetic twilight. But if we let them die or suffer when we can save or help them, we face the certainty of a moral twilight."[7]

TYPES OF GENETIC DISEASES

We can classify the two thousand or so genetic diseases into three general groups. In one category (gross chromosome abnormalities) are disorders where patients have too many or too few chromosomes in their cells, or parts of their chromosomes are misplaced. For example, people with the most common form of Down's syndrome have an extra chromosome number 21—three instead of the usual pair. Others may have an unusual number of sex chromosomes, such as Klinefelter's syndrome (XXY), Turner's syndrome (X), XXX, or XYY. By inspecting a karyotype, a visual display of the patient's chromosomes arranged in a standard pattern, a physician can readily diagnose such abnormalities. But no cure is in sight for these disorders. All the doctor can do is treat symptoms.

In the second category (polygenic disorders) are people whose chromosomes look normal but actually carry several defective genes. Environmental factors may also be important in some of these disabilities. In this potpourri are congenital heart disease, cleft palate, club feet, spina bifida, anencephaly, diabetes, and schizophrenia. Physicians can provide effective therapy for some of these disorders.

More than one hundred diseases belong to the third group—single gene defects. The list includes phenylketonuria (PKU), sickle cell anemia, galactosemia, Tay-Sachs disease, hemophilia, cystic fibrosis, Duchenne muscular dystrophy, and Huntington's chorea. Scientists have pinpointed the problem and devised specific therapy for a few (PKU, galactosemia, hemophilia). But for others, they either have not unraveled the chemical defect (Huntington's chorea, Duchenne muscular dystrophy), or they know the problem but not a solution (sickle cell anemia, Tay-Sachs disease).

But there is hope. We are discovering how to replace nature's weapon—death—with our own. Yet our swords cut both ways, helping individuals while damaging our gene pool. Now let us examine our growing arsenal.

TREATMENT

Surgery

Surgeons can often repair the structural defects from such disorders as cleft palate (malformed roof of the mouth), congenital heart disease, pyloric stenosis (narrowing of the opening between the stomach and intestine), and spina bifida (incomplete closure of the spinal canal). Although their results vary, especially with patients who have spina bifida, surgeons have helped many people with genetic disorders enter the mainstream of life. But corrective surgery does not erase their above-average risk of having children with the same defect. For example, their risks of transmitting congenital heart disease (3 percent) or pyloric stenosis (5 percent) to their children are five and twenty times higher, respectively, than normal.[8]

When a genetic disorder attacks one tissue in particular, a patient may temporarily escape by trading in his ravaged tissue for someone else's. Depending on the disorder, doctors have tried kidney, spleen, thymus, and bone marrow transplants.[9] But this usually is a stopgap, and it works for only a few genetic diseases. (We will consider transplants in more detail, including the problem of rejection, in chapter 7.)

Diet

Diet is another weapon we can use against a few genetic diseases, especially those involving a single defective gene. For example, a person with galactosemia or phenylketonuria (PKU) lacks the enzyme to metabolize galactose or phenylalanine, respectively. When a patient eats the substance he cannot metabolize, that material (and its derivatives) gradually accumulates in his body, eventually reaching toxic levels that cause severe retardation. So here the solution is obvious: The patient should avoid food with galactose or phenylalanine.

Yet it is not quite that simple. Everyone, even a person with PKU, needs some phenylalanine for growth and development. And it is not all that easy to find suitable types of food; galactose abounds in dairy products, and phenylalanine occurs in almost all proteins. Nevertheless, our dietitians have concocted special potions that work fairly well if the newborn begins his diet right away. Then when his brain is developed (about age eight), he can safely join his friends at the local hamburger palace.

Sometimes when the fetus has a genetic disease, diet therapy must begin during pregnancy. With PKU, there's no immediate danger if the mother-to-be does not have the disease herself, for her enzymes will prevent a buildup of phenylalanine. But if she does have PKU, she needs a strict diet to prevent her excess phenylalanine (which she can tolerate) from harming her fetus.

One of the most remarkable examples of nutrition therapy during pregnancy occurred in 1974. The expectant parents had lost an earlier child from a rare disease (methylmalonic acidemia), which had been diagnosed posthumously from about one-thirtieth of an ounce of the infant's urine. Laboratory tests showed they were unlucky again: Their second fetus had the same disease. Since one form of this disease responds to vitamin B_{12}, the physicians decided to inject the mother with massive doses of the vitamin—five thousand times the normal dose. It worked. Enough B_{12} reached the fetus to prevent damage. When the baby girl was born, she immediately began a strict diet and received occasional injections of B_{12}. Perhaps in time she will outgrow some of those restrictions, like people with PKU or galactosemia.[10]

Enzyme Therapy

People with a single defective gene often cannot make a needed enzyme, and this unbalances the delicate ebb and flow of their metabolism. Here we might expect that the solution would be to supply them with the enzyme they need.

This idea looks good on paper, but only recently have scientists been able to try it. One of the pioneers is Roscoe Brady of the National Institute of Neurological Diseases and Stroke. He and his coworkers have identified the missing enzyme in several different diseases, isolated the enzyme from a natural source, and then injected enzyme preparations into patients with the disease. The first hint of success came in 1973, when he tried enzyme therapy on two people with Fabry's disease, a disorder where fats accumulate in certain tissues and cells. Although the injection reduced the amount of fats in their blood to normal, the patients regained their abnormally high levels in two days.[11] One year later Brady reported that enzyme therapy for another disorder (Gaucher's disease) had helped two patients for up to six months.[12]

While those experiments showed enzyme therapy might work someday, they also revealed some problems. For one thing, people cannot take enzymes orally; our digestive tracts would destroy enzymes before they could act. And our bodies normally break down proteins and replace them with newly synthesized ones, so patients would need enzyme injections every few days for life. Yet prolonged therapy can trigger the immune system to reject the stranger moving into the neighborhood. All of these complications are familiar to diabetics who take insulin, a protein.

Although these problems are daunting, they are not necessarily insoluble. In fact, insolubility may be one solution. Someday physicians may implant, perhaps every few months, insoluble or encapsulated enzyme preparations with a timed-release action. Another idea is to trap enzymes inside fatty droplets, called liposomes, where the enzymes can work while hiding from the immune system. And scientists may be able to tailor liposomes to carry their cargo inside specific types of cells.[13] Liposomes might even protect proteins from digestion, so patients could take their enzymes orally. Indeed, this idea is being tried with insulin.[14] Yet another possibility is to redesign enzymes to help them elude the body's arsenal for degrading or rejecting them.[15]

Nevertheless, a method for delivering magic bullets would not be much good without the bullets. A patient cannot just stroll into his corner drugstore and buy a few ounces of whatever enzyme he happens to need. With one of Brady's patients, for example, it took scientists two years to learn how to isolate (from human placentas) small amounts of the needed enzyme. The cost was five hundred dollars per injection, and the supply soon ran out.

On the horizon are several possible answers. Instead of isolating enzymes from organisms, which generally have low concentrations of the desired goods, scientists may be able to manufacture specific enzymes in the laboratory. Bruce Merrifield and coworkers at the Rockefeller University developed a method, suitable for automation, which they used to synthesize a biologically active enzyme.[16] Whether this method will prove practical remains to be seen. The second option, genetic engineering, is to make microbes into miniature factories for specific enzymes. This is a truly exciting prospect that we will explore in the next chapter. We will also examine the ultimate solution for a person who lacks an enzyme: supplying him with the right gene so he can permanently make the enzyme he needs.

For now, enzyme therapy is a provocative, expensive, and unproved way to treat a few dozen genetic diseases. Physicians have not yet used it on a long-term basis with a single patient, or even a married one.

Other Methods

We have drugs that relieve the symptoms of several genetic diseases, including the Lesch-Nyhan syndrome (a severe neurological disorder) and cystinuria.[17] In time, sickle cell anemia also may join the list, for scientists are testing several compounds that prevent abnormal red blood cells from sickling.

Proteins are another form of therapy. For example, many people with diabetes take insulin. We also have protein preparations to help make blood clot. Since the injections cost many thousands of dollars a year per person, people with hemophilia can trade their physical problems for financial ones.

When a genetic disease produces toxic levels of a substance in the blood, the patient may get temporary relief from dialysis, a method for cleaning the blood of waste materials by passing it through a series of membranes. And someday those membranes may be embedded with specific enzymes that will selectively degrade toxic substances. This form of enzyme therapy would bypass rejection problems.

We will make impressive strides in helping people with genetic diseases, especially single gene defects, live more normal lives. But none of the methods we have examined can actually cure the disease, so the enormous benefits will carry a stiff price tag: an increasing genetic burden for our descendants and a rising demand for our medical resources.

How will we choose to pay the piper? Let us consider some ways we could have fewer babies with genetic defects.

GENETIC COUNSELING

Genetic counseling is a form of preventive medicine that has mushroomed in recent years, with about three hundred centers now operating in the United States. Through this service, people can learn about their chances of having children with a particular disorder, and they can also find out more about that disorder. Most clients seek genetic counseling because they suspect they have an above-average risk. Many already have had an abnormal child, or their relatives have had some disorder. Other likely candidates are women who have had two or more spontaneous abortions, and couples where the husband and wife are blood relatives.

The first step is crucial. Using whatever information his client can supply, the counselor compiles as complete a family history as possible. After eliminating nongenetic disorders, he may detect a pattern that indicates whether his client carries a particular disorder. The counselor may ask the client and other family members to have a physical examination. They also may have chemical tests, for now there are methods to diagnose carriers for over sixty disorders. After he has all the necessary information, the counselor assesses the risk factor.

Estimates for single gene defects are simple enough. If one parent has a dominant disease, there is usually a 50-percent risk for each child. If both parents carry the same recessive disease, the risk factor is 25 percent for a child having the disease, 50 percent for being a carrier, and 25 percent for neither being a carrier nor having the disease. If one parent is a carrier, none of their children will have the disease, but an average of half will be carriers. If the mother carries a disorder on one of her X chromosomes, however, each daughter has a 50-percent chance of being a carrier while each son has a 50-percent risk of having the disease. In all these situations, the risk is the same for each child, regardless of how his siblings fared. And the exact risk factors are slightly different from those mentioned; spontaneous mutations add to the risk while spontaneous abortions (which are more frequent with defective fetuses) slightly lower the odds on giving birth to a baby with a disorder.

For many genetic diseases, especially polygenic disorders, the estimates are not as clear-cut. Since the risks are often affected by other factors (for example, the sex of the child and the parents' age, race, and location), the counselor uses statistical data on how often defects occur under certain sets of conditions. As more data are compiled, the counselors will be able to refine their estimates.

After the counselor finishes his analysis, it is the clients' turn. Should they chance a pregnancy? Often it is an agonizing decision. They have to weigh the risk factor itself, the severity of the disease, the prospects for treatment, their resources, their feelings about having a child with that disability, the effect on other family members, and their reproductive alternatives. They may also consider whether the disease can be diagnosed early enough in the pregnancy to allow an abortion.

There are not any easy answers here; there are just too many variables to try to define a maximum "acceptable" level of risk. In the end, after the counselor has provided information, options, and insights into the practical reality of those options, the prospective parents must search deep within themselves for the answer.

GENETIC SCREENING

Many people who do not seek genetic counseling are unknowingly playing reproductive roulette. In order to alert them, many states and communities have established genetic screening programs for "the identification of individuals who may profit from genetic information."[18] It has been a bumpy road, as we can see by comparing two programs.

Tay-Sachs disease is a recessive condition that is uniformly fatal; the babies become blind and mentally retarded, and they die within three or four years. About one in three hundred people carries this disease, but carriers are ten times more common among mid- and eastern European Jews (Ashkenazi Jews). Physicians have no effective treatment, but they have an accurate and simple test for carriers. They also can diagnose this disease in fetuses early enough for an abortion.

In 1971 Michael Kaback and his colleagues at the Johns Hopkins School of Medicine began a screening program in the Baltimore–Washington, D.C., area for carriers of Tay-Sachs disease. The Jewish community responded enthusiastically to the voluntary program, and it soon spread to other locations. By 1976 more than 100,000 people had been tested and 4,200 were diagnosed as carriers.[19] Among them were seventy-nine couples where both members were identified as carriers. Those couples had a total of forty-six pregnancies, and prenatal diagnosis revealed that nine of those fetuses had Tay-Sachs disease. In each case, the parents chose to abort. None of their other babies have had the disease.

Sickle cell anemia is another recessive disease that affects primarily one group of people. About 9 percent of the black population in the United States are carriers, and one in six hundred black babies has the disease. People with sickle cell anemia experience episodes of excruciating pain, but a few reach middle and even old age. Although there is no effective treatment yet, the prospects are encouraging. Carriers have sickle cell trait, a harmless condition that can easily be diagnosed because some of their red blood cells have the telltale sickle shape. Only in 1976, however, did it become possible (though difficult) to diagnose sickle cell anemia in a fetus.[20]

Caught up in a wave of enthusiasm for genetic screening, several states established mandatory sickle cell screening programs in 1971 and 1972. Some required screening when blacks applied for marriage licenses. Others targeted black school children for the tests. But those programs were short-lived: In

1972 Congress essentially made it mandatory that screening programs be voluntary.

What went wrong? In a nutshell, the lawmakers did not anticipate the social impact of their programs. Many blacks viewed mandatory screening as a new type of discrimination and a not-so-subtle form of genocide. Some people identified as having sickle cell trait felt inferior. For example, they considered themselves less desirable partners for reproduction and "unsuitable" for one out of every ten black mates. Indeed, since there was not a prenatal test for sickle cell anemia, the only way high-risk couples could avoid the risk was to bear no children at all. And misinformation about sickle cell trait led to still other problems: Carriers found it more difficult to get jobs, and many had to pay higher rates for life insurance.

It was an expensive lesson, but we have learned some important things about genetic screening. For one thing, if screening programs are to do more good than harm, they must be voluntary, with the results kept confidential. In addition, the genetic counselors need to inform participants about the carrier condition, risks to their children, the severity of the disease, options for treatment, and whether there is a method of prenatal diagnosis. Above all, the counselor must be supportive and help carriers face the future positively.

We will not see an epidemic of new screening programs for carriers. Most diseases are too rare for mass screening to be practical, and for some there is not a simple, accurate, and inexpensive test. But there is another type of screening that will expand: testing newborns for genetic diseases. Here the emphasis is not on prevention, but on early diagnosis and treatment.

In most states the long arm of the law reaches out to prick the heel of each newborn, squeeze a few drops of blood onto a paper disk, and send it to a testing center for PKU. About one in thirteen thousand babies in the United States has this disease, and early diagnosis lets them start their special diets right away. Since there are a few false positive results, follow-up tests are sometimes necessary. In addition, it is important to monitor each infant's progress on the diet, for too little phenylalanine can also cause retardation. Despite these complications, though, the benefits are so compelling that we have come to accept even mandatory PKU screening. Not only do these programs safeguard the quality of individual lives, but they cost much less than maintaining the undetected, and thus untreated, PKU people in institutions for the mentally retarded.[21]

Should we also be screening infants for other diseases? Since blood samples are already taken for PKU, we could add other tests at a modest increase in cost. Indeed, in 1974 New York began a mandatory screening program for seven genetic diseases, including PKU, galactosemia, and sickle cell anemia. Other states are following suit. In fact, one lawyer has predicted:

"Within the next decade, virtually every newborn in America may be tested for a host of genetic diseases."[22]

But there are limits. For one thing, we do not have a simple and reliable test for each disease. And although no one wants to put a price on a human life, we must draw a line somewhere in terms of cost versus benefit. One of the more common diseases tested in the New York program, homocystinuria, occurs only an average of once in 160,000 births.[23] If we routinely screened all infants for very rare diseases, including those for which there is no effective therapy, the main beneficiaries would be some previously unemployed laboratory workers.

One change seems in order for any expanded infant screening programs: participation should be voluntary. It is reasonable, though, to require that parents have information about the programs and access to counseling. This arrangement might increase the cost and paper shuffling a bit, but the number of participants would not change much. Voluntary screening for PKU attracts 90- and 95-percent participation rates in Quebec and Ontario, respectively.[24]

Voluntary programs would relieve troublesome side effects of screening, such as their differential effects on racial and ethnic groups. Otherwise, for example, we would have to decide whether to require sickle cell tests on all infants (a waste of money) or just blacks (a form of discrimination). Another problem is that screening reveals pertinent (and impertinent) evidence about who the father is. The results are tainted by the fact that spontaneous mutations can produce unexpected patterns. Nevertheless, if a baby found to have sickle cell trait were born to a couple, neither of whom had sickle cell anemia or trait, the husband might want to find a quiet place to sit down and have some long, deep thoughts about life.

Voluntary programs would also give people the right not to know. This could benefit people with certain mild disorders, especially those for which there is no effective treatment. A case in point is the XYY syndrome, a condition where men have an extra Y chromosome in each of their body cells. According to some data, these men tend to be tall, slightly below average in IQ, and somewhat inclined toward aggressive behavior.[25] They comprise about 2 percent of the men in penal-mental institutions, a figure twenty times higher than the percentage of babies born with that condition.[26] But even though the overwhelming majority of these men function normally in society, there have been exaggerated claims and widespread publicity about them being predisposed to violence. In fact, at least one woman aborted her fetus because it had an XYY constitution.[27]

Should we routinely screen infants for the XYY syndrome? One program in Boston to identify and help young XYY children was halted because of political opposition. Yet early diagnosis, plus special counseling, might have helped some of those children. On the other hand, identifying them as XYY might have stigmatized them unnecessarily. And concerns about their potential for violent, antisocial behavior might have turned into self-fulfilling prophecies. On balance, then, it may indeed be better not to screen for XYY.

Still, in genetic screening we should not push the "ignorance is bliss" argument too far. Although everyone's entitled to be ignorant, no one should abuse the privilege.

PRENATAL DIAGNOSIS

Women who fear they are carrying a defective fetus often have a safety valve: Their fetus can be tested for the abnormality in question. The advantage over genetic counseling alone is that the parents know for sure whether their fetus has the disorder. And the advantage over screening newborns is that the parents have the option of an abortion.

Most prenatal diagnoses are done by amniocentesis. Here the physician inserts a syringe through the abdominal wall of the pregnant woman and collects about twenty cubic centimeters of amniotic fluid. From that fluid, which carries a few fetal cells and waste materials, the physician must coax the critical information about the fetus. Abnormal amounts and types of waste materials occasionally give clues, but the fetal cells are the physician's best friend. Since there are not enough cells in the amniotic sample to do the tests directly, he must first multiply them in a culture medium. This takes about three weeks if he is testing for gross chromosome abnormalities and up to five weeks for metabolic disorders. The wait seems even longer to the prospective parents.

With amniocentesis, physicians can diagnose more than seventy disorders, and the number is growing unsteadily. Some familiar examples are Down's syndrome, Turner's syndrome, Klinefelter's syndrome, Tay-Sachs disease, and galactosemia. Chromosome analysis also reveals the sex of the fetus. Missing from the list, however, are relatively common diseases such as cystic fibrosis, hemophilia, Duchenne muscular dystrophy, and diabetes.

And there are a few other items in the physician's bag. Ultrasonic vibrations, a type of sonar that produces images of the fetus and placenta, can help the physician find a safe place to insert the needle during amniocentesis. Those images also reveal structural defects, including anencephaly (absence of a brain and spinal cord), spina bifida, and certain heart and kidney disorders. In addition, ultrasound shows the number of fetuses.

Another tool for detecting structural defects is the fetuscope, a sort of miniature telescope with a lens mounted in a needle. The doctor can insert this device through the cervix and then peer through the amniotic soup to see the fetus. The fetuscope may prove especially valuable in combination with amniocentesis. If the physician could see the fetus well enough, for example, he could remove enough cells to do the diagnoses directly, thus eliminating the three-to-five-week culture time. He also could remove a blood sample directly from the fetus, paving the way for the prenatal diagnosis of blood diseases such as sickle cell anemia. In fact, this has already been done successfully.[28]

The mother herself may supply crucial clues that all is not well with her fetus. Since she is the disposal unit for fetal wastes, her blood may carry signs of trouble. The most useful results so far are in detecting fetuses with spina bifida and anencephaly. Indeed, blood tests on nineteen thousand women during their sixteenth to eighteenth week of pregnancy showed a two-to-three-fold elevation in a certain substance (alpha-fetoprotein) when their fetuses had these disorders.[29] So routine blood screening for alpha-fetoprotein could identify women who should have further tests (such as ultrasound) to confirm the diagnosis. And the cost of screening everyone would be less than that of caring for the spina bifida and anencephalic children that otherwise would be born.

SOME CONCERNS

Restrictions on Prenatal Diagnosis

How thoroughly should a physician test each fetus for abnormalities? One factor he must weigh is the safety of the procedures themselves. Ultrasound appears to be safe, especially in comparison with X rays. The fetuscope, however, is only in an experimental stage. It can cause hemorrhaging and discomfort in the mother, and it can pierce the placenta. Indeed, the fetuscope has caused at least one abortion.[30]

Amniocentesis also poses a few risks. Dangers to the mother—infection and hemorrhaging—are minimal, but mixing fetal and maternal blood could sensitize an Rh-negative mother against an Rh-positive fetus. Puncturing the fetus is another concern. Although a physician can do amniocentesis as early as the thirteenth or fourteenth week of pregnancy, the chances of puncture are a bit higher before about the sixteenth week because the smaller volume of amniotic fluid leaves less margin for error in positioning the syringe. The effects of a puncture usually are temporary, but there have been several reports of permanent damage, including one fetus that lost the use of an eye.[31] Nevertheless, all these risks are now minimal because physicians are more experienced with amniocentesis and many are using ultrasound first to locate the fetus and placenta.

The greatest safety concern is that amniocentesis might trigger abortions.

In 1975 the National Institute of Child Health and Mental Development compiled data for 1,040 women who had amniocentesis and 992 who did not. They found no statistical evidence that amniocentesis caused miscarriages, premature births, birth defects, or abnormal development during the first year after birth. The study also revealed that the diagnoses were 99.4 percent accurate.[32] That makes amniocentesis one of the more accurate diagnostic procedures in medicine, though it gives scant comfort to the few individuals who, acting upon a misdiagnosis, aborted a normal fetus or proceded to have an abnormal baby.[33]

Amniocentesis is clearly safe and accurate enough to use when there is a sufficient risk—perhaps 1 percent or more—of a seriously abnormal fetus. But where should we draw the line? John Littlefield of Harvard University has predicted: "The day may come, 10 or 20 years from now, when amniocentesis will be performed in every pregnancy."[34]

That seems unlikely. For one thing, there is little reason to do amniocentesis if we lack a method to diagnose the disease in question. Furthermore, the risk of the procedure, while very small, cannot be zero. Indeed, some physicians have considered the risk worth taking only when the mother decides in advance to abort if her fetus is found to be defective. But that attitude is changing, for physicians now realize the dangers of amniocentesis are considerably less than they had thought. So perhaps the major limit will simply be dollars and sense. Amniocentesis for every pregnancy would take an immense commitment of our medical resources, and those resources are limited. It would be a bit difficult to justify doing amniocentesis on every customer who walked through the door with the requisite number of box tops.

Selective Abortion

When prenatal diagnosis reveals an abnormal fetus, the parents often seek an abortion. Since amniocentesis is not safe until at least the fourteenth week, and it takes several more weeks to do the diagnosis, a fetus often is twenty weeks or older at the time of abortion. That makes the abortion more difficult—physically, legally, and morally.

In 1973 the United States Supreme Court ruled that states could not interfere with abortions during the first trimester, but they could protect fetuses that reach the age of viability (about twenty-four to twenty-eight weeks). Since states do not allow abortion on demand once fetuses become viable, the parents must have amniocentesis done promptly if they want to have abortion as a legal option. Yet that narrow time margin could shrink further as our technology shortens the age of viability. Indeed, if the limit for viability (and abortion on demand) became much earlier than twenty-four weeks, prospective parents could not use amniocentesis for abortion decisions except in the specific situations (perhaps, for example, fetuses with serious and untreatable defects) that their state law allowed.

Now let us turn to the moral issues. Amniocentesis has expanded the use of abortion, enabling parents to control not only the number of children, but also the type. Because of this safety net, a high-risk couple is more likely to take a chance with a pregnancy, hoping for the best but knowing they can abort if their fetus has the disorder they fear. As we saw earlier, couples where both members carried Tay-Sachs disease have had dozens of pregnancies. Indeed, one woman reportedly aborted three successive fetuses with Tay-Sachs disease before giving birth to a normal child.[35]

Yet amniocentesis has had little impact on the total number of abortions. For one thing, only a tiny fraction of the people having abortions base their decision on the results from amniocentesis. Furthermore, amniocentesis sometimes has the effect of preventing abortions. Since at least 95 percent of the diagnoses reveal no abnormality, some women who otherwise would not have risked bearing a defective baby were encouraged to proceed with their pregnancies.

Selective abortion, however, brings us to the question of how much abnormality, if any, justifies an abortion. Legally, the parents must make that decision. But on what basis should they decide? In terms of morality, we can only suggest general criteria to serve as guidelines.

The prospective parents have many factors to weigh. For example, if they are worried about their fetus having Duchenne muscular dystrophy (which is carried on the X chromosome), amniocentesis can give them only a partial diagnosis. Amniocentesis will reveal the sex of the fetus, but there is no reliable test for the disease itself. So a couple may have to face an abortion decision knowing only that they have a male fetus with a 50-percent chance of muscular dystrophy.

Another complication is the wide range in abnormality between and within diseases. For example, many woman with an XXX constitution are normal physically, but they have about twice the usual risk of being admitted to a hospital for mental illness.[36] The XYY condition is another mild disorder. At the other end of the spectrum is anencephaly, where the baby lacks a brain and spinal cord. Somewhere in between is Down's syndrome, the most common condition for doing amniocentesis. Those people are mentally retarded, with IQ values averaging about 50, but with a wide range in abilities. Most are trainable, and some have IQ scores of 70 or more.

The parents-to-be must also weigh the prospects for therapy. Many disorders are untreatable. But for others, such as PKU, galactosemia, and certain malformations, the prognosis is good. Although lengthy or expensive therapy may be a burden to the other family members, it would be difficult, ethically, to justify that this outweighs the chance for the fetus to become a reasonably normal person. In some "treatable" diseases, however, the results vary considerably. For example, some children with spina bifida can function well after surgery; others are left paralyzed and retarded. Indeed, British physicians have developed medical criteria for when it is advisable to operate on newborns with spina bifida.[37] The actual decision, however, lies with the parents.

Conscious thinking ability is a basic quality of human beings, so prospective parents also should consider the mental potential of their fetus. Fetuses with certain disorders have no realistic chance of becoming full human beings. But where should we draw the line? Joseph Fletcher has suggested that an individual with an IQ of less than 20 is not a person, and someone with an IQ of 20 to 40 is only marginally a person.[38] Although IQ is an arguable yardstick for mental capacity, and it cannot be measured in a fetus, some minimum threshold of intelligence (or potential intelligence) seems a reasonable criterion for "personhood." For without the capacity for self-reflection and relationships with others, an individual cannot truly enter the marketplace of human affairs.

These criteria would not justify using abortion simply for sex selection. But this application is legal and feasible right now. And some people would like to use it. In 1975 a research group at Indiana University Medical Center developed a method for collecting fetal cells by inserting a cotton swab deep into the cervix.[39] They reported that their method was 86 percent accurate in diagnosing the sex of the fetus, and it could be used as early as the ninth week of pregnancy. One member of the group reported: "A few people found out about it in some way and a couple of women called up, because if it wasn't the desired sex they wanted an abortion."[40]

Would people actually use abortion for sex selection? They already have. According to one report from China, a method similar to that of the Indiana group is 94 percent accurate and has been used to screen pregnancies. Of the first one hundred women screened, 63 percent of the women who had female fetuses chose to abort, while only 2 percent of those carrying a male fetus aborted.[41]

What's a poor (or rich) physician to do? According to one poll, 95 percent of the doctors oppose using amniocentesis plus abortion solely for sex selection.[42] To look at it another way, though, 5 percent do not oppose the practice. A similar dilemma confronts the geneticist who is asked to diagnose the sex of the fetus. May he, or the physician, deny the woman the opportunity for informed free choice if he personally believes the use of that information is immoral? According to M. Neil Macintyre of Case Western Reserve University, such people "are trapped between choosing between being unethically moral and ethically unmoral."[43]

Now let us return to the question of where we should draw the line on aborting abnormal fetuses. The clearest answers would be: Abortion is always acceptable because the fetus is not a human being; or, abortion is always wrong because it is the taking of an innocent human life. The first view is disturbing because of the great respect we should hold for a form of human life and at least a potential human being. Aborting a fetus solely because it is the "wrong" sex shows a callous disregard for the precious gift of life.

The second answer is also disturbing, for it implies that life itself is sancrosanct, that it is the ultimate good to which everything else is subordinate. But that doesn't fit with most religious traditions. Indeed, Pius XII said: "Life, death, all temporal activities are in fact subordinated to spiritual ends."[44] And according to Richard McCormick of the Kennedy Center for the Study of Reproduction and Bioethics at Georgetown University: "Life is a value to be preserved only insofar as it contains some potentiality for human relationships."[45]

If we accept either or both of these ideas, we must conclude that life is not an absolute good. Although life is exceedingly important, its value lies not in itself but in the fact that it is the framework that enables us to think, to grow mentally and spiritually, and to enter into relationships with others. Without that ability, or potential, life ceases to be meaningful. Consider, for example, an anencephalic fetus or baby, which has no potential to rise above the status of a human "vegetable." Perpetuating his mechanical life, at great psychological and physical expense and for no discernible benefit to anyone, would be an empty offering on the altar of life as the ultimate good—indeed, of life itself as our God.

Yet if we reject the two (too) simple, all-purpose answers on abortion, we are left with the unsurprising conclusion that there are certain situations where abortion is justified and others where it is not. And the best we can do here is to develop guidelines within some ethical framework that may assist the people who actually have to make the decisions.

It is just as well to leave it there anyway. Each of us can sit back in an easy chair and philosophize about what is or is not morally correct. But while there is something to be said for detachment and objectivity, talk can be very cheap if the person doing the philosophizing is not personally responsible for those excruciating decisions. According to Richard McCormick: "Moral theologians . . . can easily be insensitive to the moral relevance of the raw experience, of the conflicting tensions and concerns provoked through direct cradleside contact with human events and persons."[46]

Yes, they can. But so can the rest of us.

The Gene Pool Problem
Genetic counseling, genetic screening, and prenatal diagnosis all will help

parents have fewer defective babies. But if we expect these methods to help stem the rising tide of abnormal genes in our gene pool, we are likely to be disappointed.

Genetic counseling and genetic screening of adults might deter some high-risk couples from the gambling tables. But amniocentesis will lure many of them back. Still, we might expect that amniocentesis plus selective abortion would improve the quality of our gene pool. Yet it's not that simple. The problem comes if the members of a couple arrange their affairs (or reproduce with each other) to have a specific number of children. In the case where both parents carry the same disorder, each defective fetus they abort would be replaced by a child without the disorder, but with a two-thirds chance of being a carrier. And carriers are more likely to grow up, marry, and pass their defective genes on to their children. According to several calculations, this "reproductive compensation" would cause a small contamination of our gene pool.[47] And the downhill slide would be worse for diseases carried on the X chromosome if couples at risk aborted males and replaced them with girls.

So we find that genetic counseling, genetic screening, and prenatal diagnosis are not likely to reduce the load of defects in our gene pool. Unfortunately, what is good for individuals today could be harmful to future people. This means we must look elsewhere for a way to defuse our genetic time bomb.

One option is the head-in-sand approach. We could leave it to future generations to worry about. Or we could simply write off the problem as a necessary cost of a civilized society where we help the less fit survive. We might, for example, point to a study showing that monkeys have a tenfold lower incidence of abnormal babies than we do; perhaps this is a price we must pay for our humanity.[48] We also could sit back and let the pattern run its course. When our genetic load overpowered our medical resources, natural selection would again bare its fangs, culling out the growing numbers of unfit people. That is a pretty ugly "solution," though. Perhaps we can do better.

Another option is to let natural selection take its toll in situations where the defect is especially severe. On a small scale, we are already doing this with infants. One example, which received considerable publicity several years ago, was a newborn with Down's syndrome who also had an intestinal obstruction. Because the parents felt their baby would cause severe problems for themselves and the other children in the family, they did not authorize surgery to correct the obstruction. The infant could not digest food, and he slowly starved. Death came fifteen days after birth.[49]

Selective infanticide is becoming more common, or at least its practice is now more open. Between 1970 and 1972, forty-three babies died at the

Yale–New Haven Hospital after special treatment was discontinued; in each case the parents and physicians together concluded that the infant had a negligible chance for "meaningful life."[50] Indeed, Joan Hodgman of the University of Southern California School of Medicine remarked: "If we have a baby that I know is malformed beyond hope, I make no attempt to preserve life."[51]

Infanticide is passive; the health personnel withhold treatment, letting nature take its course, but they do nothing directly to cause death. This helps protect them against a charge of murder. Another advantage is psychological: The participants may feel "nature" is responsible for the death. Nevertheless, the moral distinction between active and passive infanticide lies in the eye of some other beholder. In both situations the participants deliberately select and carry out measures that they expect will hasten the death of the infant. *How* it is done is important, but not as important, ethically, as the fact that it *is* done. For example, once the decision was made not to save the baby with Down's syndrome, it might have been more humane to kill him directly than to let him slowly starve to death.

For several reasons selective infanticide will not do much to improve our gene pool. For one thing, moral objections will rightly limit this practice to a very small number of babies, most of which would not have reproduced anyway. Yet even if there were no moral constraints, and we had widespread infanticide, we still would not relieve the problem; for the genetic effect would be like that of amniocentesis plus selective abortion—fewer children with disorders, but more children who are carriers.

Option three is to treat the hereditarily ill but restrict their reproduction. Genetic counseling would be the keystone of a voluntary program, but amniocentesis would undermine its effect on the gene pool. Yet even without prenatal diagnosis, some people would play reproductive roulette. According to genetic counselors: "Couples . . . knowing that they have one chance in four of having a seriously defective child, and that two out of four of their children are likely to be carriers, still frequently take a chance that things will turn out all right."[52]

Since voluntary abstinence is not likely to sweep the world, we might consider stronger measures. But how far could we go? Many people believe God joined sexual intercourse with procreation, and therefore it would be immoral for man to separate them. And many regard reproduction as a basic human right. According to the United Nations' Universal Declaration on Human Rights, every married couple has a right to "found a family."[53]

On the other hand, we must question the notion that procreation is an absolute right. It is hard to accept the implication that parents own children and have unlimited rights to produce them even if they are likely to be born with serious and untreatable defects. According to Paul Ramsey, if a person knows he is likely to produce a seriously defective child, "then such a person's 'right to have children' becomes his duty not to do so, or to have fewer children than he might want (since he never had *any* right to have children simply for his own sake)."[54]

We also could argue that society has a right to restrict the reproductive freedom of high-risk couples unless they bore the cost of rearing their defective children. For example, it costs an average of $250,000 for society to provide a lifetime of care for each person institutionalized with Down's syndrome.[55]

Still, we would have a hard time implementing an effective but fair program. Suppose we decided that people should not reproduce if their genetic risk were fairly high (say 10 percent or more), the disease was serious and untreatable, and there was no method for prenatal diagnosis; and if there were a prenatal test, we would require them to have amniocentesis. How would we enforce those rules? One suggestion is to withhold health insurance payments from high-risk couples who knowingly proceeded to have children with serious defects.[56] Yet that would penalize the children as well as their parents, and it would affect only the people who have insurance. Furthermore, the main effect, if any, would be to discourage people from seeking genetic counseling; for some would prefer the risk of ignorance to the risk of no insurance payments. Besides, strict enforcement of this or any other measure might trigger a wave of respectability for an age-old genetic option: adultery.

The direct approach would be compulsory sterilization, a method that has been used for both genetic and social reasons. In 1965, for example, two thousand Danes were sterilized under their law dealing with the retarded and certain criminals and psychotics.[57] Many states also permit involuntary sterilization under certain conditions, but the actual practice is rare.

We might regard sterilization as the price of admission into a civilized society, but we would be hard pressed to justify its widespread use. For one thing, there is an obvious potential for abuse. (Suppose, for example, we decided to sterilize everyone with sickle cell anemia.) Yet sterilizing a few people, some of whom would not reproduce anyway, would make only a small dent in our gene pool problem. In order for our program to be truly effective, we would have to sterilize all the carriers as well. The only trouble is that in a few years no one would be left to applaud our solution.

A fourth option is to improve our gene pool by selective breeding. First, we would choose the "superior" qualities we wanted. Then, if we wanted the maximum effect, we would classify each person; only those with Grade A qualities could reproduce, and only with each other. Artificial insemination would be a handy way to spread the good genes around, and we would encourage the blue-ribbon people to produce as many children as they could.

Each child would in turn be graded, and some would qualify to become parents for the next generation. Eventually we would amass a glittering gene pool that ensured our production of superior people.

Still, there is a problem or two. For a truly effective program, a powerful group would have to get large numbers of people to "cooperate." And that group probably would select the "superior" genes. Then they would make the startling discovery that they were the very ones who possessed all those outstanding qualities. So, as members of the elite group, they and their descendants would be obliged to accept the permanent burden of governing the genetic also-rans.

There would be nonpolitical problems as well. For one thing, selective breeding on a major scale would undermine our genetic diversity. And many desirable qualities could not be chosen anyway, for some do not come wrapped in genetic packages. Another problem is that breeding for certain genetic qualities might unintentionally select for other qualities that are not so desirable. People of high intelligence, for example, may have a propensity for postnasal drip.

There are, of course, less extreme measures. Two eminent biologists, H. J. Muller and Julian Huxley, advocated that we make available (by AID) the sperm from people with distinguished genetic features. Yet even if we could decide who those sperm donors should be, and they agreed to donate, we would see little improvement in our gene pool unless there was a heavy demand for their samples.

John Maynard Smith, a British geneticist, calculated the effect for one version of this plan. He assumed AID was used to increase IQ, and 1 percent of all women agreed to have half their children by this method. He also assumed the husbands of those women were a random sample of the population, and the average IQ of the sperm donors was 15 points above average. Smith concluded that the effect of this program would be to boost the average IQ score of the next generation by 0.04 point.[58] It is a bit hard to work up much enthusiasm for a figure like that.

A CLOSING THOUGHT

As nature plays her genetic games, she challenges us to respond. And our answers tell us much about ourselves. At the moment, we are moving away from an ethic of the absolute sanctity of life and toward a greater emphasis on the quality of life. Indeed, Joseph Fletcher has said, "The life sciences have made QOL (quality of life) the Number One moral imperative of mankind."[59] And many people would agree with the view of Bentley Glass: "No parents will in that future time have a right to burden society with a malformed or a mentally incompetent child. Just as every child must have the right to full educational opportunity and a sound nutrition, so every child has the inalienable right to a sound heritage."[60]

I cannot imagine anyone not wanting each child to have a completely sound genetic heritage. But we live in the real world, so we must decide how we will cope with imperfection. There are degrees of incompetence and malformation, and it is not at all obvious that abortion is the right answer for all imperfect fetuses.

This brings us to a crucial point. Just as the sanctity of life has been held too strongly as an absolute, uncompromisable good, we must beware of the pendulum swinging too far in the other direction. We should not presume that "normality," a term none of us can define adequately, is a prerequisite to a full and meaningful life. People are born with an enormous range of abilities and disabilities, and many "defective" people have not only experienced rich lives, but they also have made life richer for the rest of us.

Perhaps physical perfection is incompatible with mental and spiritual perfection. We lose a sense of compassion, caring, and love if we exercise these qualities only when it is easy. Indeed, if we come to view defective people as unwanted intrusions who diminish the quality of our lives, the greatest pity will not be what is happening to them. It will be what has happened to us.

Today there is much talk about the possibility of human genetic modification—of designed genetic change, specifically of mankind. . . . I think this possibility, which we now glimpse only in fragmented outline, is potentially one of the most important concepts to arise in the history of mankind. I can think of none with greater long-range implications for the future of our species. Indeed this concept marks a turning point in the whole evolution of life. For the first time in all time a living creature understands its origin and can undertake to design its future.

Robert Sinsheimer

We have met the enemy and he is us.

Pogo by Walt Kelly

6

Genetic Engineering

For over three billion years nature has controlled our genetic destiny. Her rules were simple but effective: Individuals with new genetic features arose by chance (through mutations and the random assorting of genes during sexual reproduction), and those with the best reproductive ability contributed most to the gene pool of the next generation. Over an enormous length of time some startling changes occurred, including the appearance of a remarkable animal that could think rationally and abstractly.

Now that animal, the culmination of nature's slow but steady process, is starting to change the rules. He is challenging the idea that he must take what he gets from nature, that her misfits must simply fall by the wayside. When he discovered the stuff of life—DNA—and learned how to manipulate it, he suddenly opened the door to a powerful new arrangement. Now he can modify, and perhaps improve, the genetic combinations nature provides; no longer is he a captive of the old arrangement.

Genetic engineering is a way to redesign organisms, including us. It offers us a way to escape a genetic doomsday, to produce more food, to fight disease, and to create new organisms our world has never seen before. According to Kimball Atwood of the University of Illinois: "We could, for example, produce an organism that combines the happy qualities of animals and plants, such as one with a large brain so that it can indulge in philosophy and also a photosynthetic area on its back so that it would not have to eat. It is not inconceivable that there could be humanoids with chlorophyll under their skins so that they would look like the enormous green man on a can of peas."[1]

The prospects boggle the mind. Indeed, the effects we can imagine range everywhere from utopia to the destruction of man as we know him. But when we return to reality, we discover there is a great gap between what is theoretically possible and what is within our grasp. Before we can make that distinction, however, we must understand the state of the art. So let us begin by examining five ways to supply organisms with new genes.

THE TECHNOLOGY

Mutations

Nature uses mutations to make new genetic blueprints, and we could try the same tactic. Indeed, we know several ways to change the chemical form of the basic building blocks (nucleotides) in DNA. The trouble is that most of nature's experiments fail, and the mutations fail to survive to reproduce their kind. We would face the same problem. With microbes, plants, and simple animals, we might be willing to accept large numbers of defective individuals for the chance of making a few with the features we wanted. In fact, we do just that when we use radiation to develop new plant strains. But we would hardly pay this price for the sake of reengineering humans.

Mutations could be a way, in theory, to correct genetic defects. People with certain disorders have a single defective gene, and that gene may have just one nucleotide in the wrong chemical form. The defect arose earlier in their genetic history because of a mutation, and we could cure it if we could mutate the abnormal nucleotide back to its original form.

The idea is simple, but only on paper. Our methods for inducing mutations (radiation, chemical treatments) are too random to make the precise change we would need. One problem is that there are only four different kinds of nucleotides in DNA, so the genetic landscape is monotonous. And with about three billion nucleotide pairs in human DNA,[2] forty-six chromosomes, forty thousand to one hundred thousand different genes,[3] and up to thousands of nucleotides in a gene, the chances of finding and correcting just one particular nucleotide in one particular gene on one particular chromosome would be like trying to find a preselected snowflake in a blizzard. Except it would be much worse. For a mistake—inducing a mutation in the wrong place—could be disastrous.

In short, the prospect for using mutations to redesign humans, or to correct their genetic defects, is approximately zero.

Cell Fusion

Another way to give cells a new genetic blueprint is to fuse them with cells that have the desired genes. In 1965 Henry Harris and his coworkers at Oxford University discovered that certain inactivated viruses will cause cells to fuse together, forming hybrid cells with a single nucleus that contains DNA from both of the original cells.[4] In culture those cells can multiply, divide, and do beginning calculus.

They also learn other new tricks. When cells from the same species fuse, the hybrid cells usually have about twice the normal number of chromosomes, and they show genetic features from each of the original cells. For example, when human cells lacking a particular enzyme were fused with human cells lacking a different enzyme, the hybrid cells synthesized both enzymes.[5] But when cells from two different species fuse, a different pattern emerges. A strange sort of genetic war takes place, leaving the hybrid nucleus with all the chromosomes from one species and only a few, or none, from the other. One humbling discovery is that when man and mouse cells fuse, it is the mouse chromosomes that prevail.[6] Scientists have even fused human cells with those of the mighty mosquito, but in that study, perhaps fortunately, they did not determine which set of chromosomes triumphed.[7] Yet even when the chromosomes of one species are pulverized, a few of their genes may survive. In one experiment, for example, hybrid mouse-chicken cells appeared to have only mouse chromosomes, but they synthesized a chicken-type enzyme, even after one hundred generations in culture.[8]

Since fertilization is a form of fusion, we might expect that sperm cells would be ideal for carrying new genes into cells. But body cells have not proved as hospitable as unfertilized eggs. Although scientists have tried to fuse sperm with body cells, so far the sperm have not managed to deliver their genetic cargo into the nucleus.[9]

Supplying new genes to a cell is not the only way fusion can cause genetic changes. The other way is to change how a cell uses the genetic information it already has. Through a complex switchboard of "on" and "off" signals, a cell normally uses only a small percentage (perhaps 5 percent) of its vast storehouse of genetic information. In some experiments, however, cell fusion has activated the "on" switch for some previously repressed genes, thus causing the cell to make greater use of its genetic potential.[10]

So we can see that fusion is a powerful way to make genetic changes. The problem—as far a genetic engineering is concerned—is that fusion is a crude way to make precise changes. Outfitting a cell with chunks of genetic material from another cell will rarely produce exactly the desired change. It is a bit like trying to repair a television set blindfolded, with ear plugs, using plumbers' tools, and wearing boxing gloves.

But there is a catch: With fusion, the right result literally one time in a million might be good enough. For sometimes scientists can use special diets to selectively starve the hybrid cells they do not want. For example, cells from galactosemic people cannot metabolize galactose, so they will not grow in a culture medium where galactose is the only sugar. But if those deficient cells were fused with cells carrying the necessary genes, and then grown on that culture medium, only the hybrid cells that could metabolize galactose would survive.

Overall, fusion is a very useful tool for basic research, and it is of modest value for tailoring plants and microbes. But using it directly to reengineer man is strictly a long shot.

Recombinant DNA

For several decades scientists have known that bacteria will take up and use naked strands of DNA. More recently, they discovered that plant and animal cells can do the same thing. For example, Allen Fox and his coworkers at the University of Wisconsin have transferred at least eleven new genes into flies by soaking the eggs from which they hatched in a broth of DNA. [11]

Cells from mammals, including man, also take up pieces of DNA. It is a hit-or-miss proposition, though, for the DNA quickly learns its new host is as hospitable as a piranha. But in 1973 scientists discovered there is much less damage to the new DNA when it is supplied as whole, purified chromosomes.[12] With this method they were able to insert human genes into mouse cells, producing cells with a permanent ability to make a human-type enzyme.[13]

About the same time, a powerful new method—called recombinant DNA—suddenly captured the attention of the scientific world. In 1973 Stanley Cohen and Annie Chang (both at Stanford University) with Herbert Boyer and Robert Helling (both at the University of California, San Francisco) published a method for supplying genes to cells.[14] Drawing on the work of other scientists, they used a special vehicle—a plasmid—to deliver the new genes, and they used a remarkable new technique for splicing the desired genes into the plasmids.

First, let us examine the vehicle. Plasmids are small ringlets of DNA that inhabit bacteria but are not part of their chromosomes. They supply optional bits of genetic information, such as resistance to certain antibiotics. One reason plasmids are so handy for genetic engineering is that bacteria readily accept them and use their genetic information. In fact, bacteria can naturally pass plasmids to each other, thus giving their colleagues new abilities. But now scientists have joined the game, for they have learned how to isolate plasmids and insert them into other bacteria.

The second step in the new method—splicing genes into plasmids before putting them in bacteria—is what makes recombinant DNA such a powerful technique, for this means genetic changes can be made with great precision. Gene splicing is done with a scalpel called restriction enzyme, which cuts DNA in very specific places, producing fragments with "sticky ends." When those fragments are mixed together in the presence of some glue (an enzyme called DNA ligase), they join end to end in various combinations. (This is the actual recombinant DNA material.)

So now the recipe for genetic engineering is to cut open plasmids with a restriction enzyme, do the same to the DNA from some other cell, and then mix both sets of fragments together. The pieces will seal to form hybrid plasmids of various compositions. The microbes will engulf those plasmids, and some of them will suddenly learn new tricks, like making a useful protein. This may be just enough of an edge for them to survive in a culture medium designed to eliminate their underendowed brothers.

Gene splicing is not just a theory. Scientists have used hybrid plasmids to supply bacteria with DNA from insects, sea urchins, frogs, yeast, mammals, and unrelated bacteria.[15] And the microbes actually use that new DNA. But we do not know how effective this method will be in redesigning plant and animal cells, which apparently do not contain plasmids. Scientists have learned, however, that plant cells can take up bacterial plasmids; and there are a few reports of plasmidlike structures in certain animal cells.[16]

Genetic engineering, by whatever method, would be more precise if scientists could insert specific genes instead of assorted chunks of DNA. But first they would have to have those single genes, or small groups of genes, on hand. Could they do that someday? They already have—three different ways.

The first solution came in 1969, when Jon Beckwith and his coworkers at Harvard University were able to isolate a small group of genes from a bacterium.[17] The second method surfaced one year later, when scientists at the University of Wisconsin, led by H. G. Khorana, chemically synthesized a yeast gene.[18] But because it lacked the "start" and "stop" signals (which enable the cell to translate the gene's information), that man-made gene could not work inside a cell. Seven years later, Khorana's group (then at the Massachusetts Institute of Technology) synthesized a bacterial gene, complete with the necessary signals, and inserted it into a bacterium. It worked.[19] The project took nine years and twenty-four postdoctoral fellows to complete, but it uncovered new methods that will simplify such projects in the future. (Fortunately, too, graduate students and postdoctoral fellows are an endless source of cheap labor.)

The third way to make a gene is to reverse what the cell does. Cells normally use their DNA to make a type of RNA (messenger RNA), but there is an enzyme that does the reverse: From a specific messenger RNA, the enzyme synthesizes the DNA coding for that messenger RNA. So if a scientist can isolate the messenger RNA corresponding to a particular gene, he can make the gene. Indeed, scientists have used this method to make a rabbit gene for hemoglobin and a rat gene for insulin.[20] But when those genes were spliced into plasmids and put in bacteria, they failed to function. One reason is that they lacked the "start" and "stop" signals. So with this method, scientists will have to identify those signals, synthesize them, and attach them to the genes before putting them in plasmids.

Viruses

Since cells often give foreign DNA a rude reception, the visitor needs an escort—a sort of Trojan horse—that knows its way around in the cell. We have already examined one such candidate—a plasmid. The other possibility is a virus.

Viruses are strands of DNA or RNA wearing a protein coat. We can think of them as syringes that inject only nucleic acid into the host cell. That viral nucleic acid may become a permanent part of the cell's genetic machinery. Or it may cut out pieces of the bacterial DNA and bring them along when it infects another cell. Lewis Thomas has written in *The Lives of a Cell*: "We live in a dancing matrix of viruses; they dart, rather like bees, from organism to organism, . . . tugging along pieces of this genome, strings from that, transplanting grafts of DNA, passing around heredity as though at a great party."[21]

For genetic engineering, then, the idea is to find a virus that will insert the desired gene(s) into the target cells without harming them. If the virus naturally carried those genes, it would be simpler. If not, and this would usually be the case, scientists first would have to attach the genes to the virus. Having pure genes, natural or synthetic, would make things easier.

Can it be done? It already has. In 1976 Ronald Davis and his research group at Stanford University used a virus to carry some yeast genes into bacteria. They treated the viral DNA and yeast DNA with a restriction enzyme, sealed together various combinations of the fragments, and infected bacteria with that hybrid DNA. Some of the yeast genes functioned in their new home.[22]

A great advantage of viruses is that they are a way to reengineer not only bacteria, but also plants and animals. For example, viruses have been used to transfer bacterial genes into tomato cells, a rabbit gene into monkey cells, and bacterial genes into cells from a person with galactosemia. In the latter experiment, the virus first infected some bacteria, pirating away their genes for galactose metabolism, and then brought those genes along when they infected the human cells in culture. A few of those cells suddenly began making the galactose-metabolizing enzymes, and they passed that ability on when they reproduced. In short, the genetic defect in those human cells was corrected.[23]

Chimeras

According to Greek mythology, the Chimera was a fire-breathing she-monster with the head of a lion, the tail of a serpent, and the body of a goat. We have also come to use *chimera* to describe plants and animals containing two or more genetic types of cells. One example is a person who receives a transplant from someone who is not his identical twin.

Chimeras may arise naturally when one of their cells mutates and establishes a different genetic line within their body. This is also called a mosaic effect, and there are several examples in animals, including man.[24] Among the most striking are sheep fleece mosaics, animals that have patchy, underdeveloped regions in their coat of wool. The proportion that is deficient varies considerably, and it probably depends on when in development the telltale mutation occurred; the earlier the mutation, the greater the effect.[25]

Deliberately making chimeras or mosaics would be a type of genetic engineering. Perhaps we could provide animals with cells carrying genes that are superior to those they already have. Yet our experience with transplants points to a likely problem with this approach—rejection. It might be possible, however, to make the new cells so similar to the originals that the body would not recognize them as "foreign." Here the idea would be to take cells out of the body, try to change only one or two genes, and then reimplant them.

Another solution is to supply the other cells early in development, before the immune system learns what is "self" and what is "foreign." Beatrice Mintz and her coworkers at the Institute for Cancer Research in Philadelphia have pioneered some remarkable methods for doing this. They can inject one or more cells of a different genetic type into a young embryo, or simply combine two (or more) embryos into a single embryo, and then implant the remodeled embryos in suitable females. The offspring have immune systems that reject "foreign" substances, but tolerate both genetic types in their own tissues.[26]

Scientists have used these methods to produce thousands of chimeric mice and a few chimeric rabbits, rats, and sheep.[27] By mixing embryos from different types of mice—for example, albino mice and those with black coats—they produce animals whose coats are a patchwork of black and white. But scientists have had little success in producing chimeras from two different species. Although they have developed several types of chimeral plants,[28] their between-species experiments with animals have not produced proven chimeras that survived beyond birth. A major reason is that embryos from different species march to different drummers, each following its own rate and pattern of development.[29]

But in 1978 came a modest breakthrough. A team of scientists working at the Jackson Laboratory in Bar Harbor, Maine, produced three mice that contained in some of their organs more than 2 percent human DNA. Developed from mouse embryos that had been injected with hybrid mouse-human cells, the brave new mice looked perfectly normal. Only their geneticist knew for sure.[30]

These methods are invaluable research tools, but we would find them a bit complicated to use routinely on humans. To make the system work, we would have to isolate each embryo, analyze its genetic fitness (if we wanted to correct disorders), have on hand a batch of embryonic cells with the desired features, mix them into each deficient embryo, and then implant the reengineered embryos in women or develop them in artificial wombs. And even after all that, we would not be sure—at least not with our present level of knowledge—what proportion of the "better" cells the baby would have. So the list of people waiting to use this technology is rather short at the moment.

We have seen that the genetic engineers have several methods at their disposal. When they redesign bacteria, they usually use the recombinant DNA method with plasmids, though viruses also work. Viruses, fusion, and plasmids all are possible tools for developing new plant strains. But genetic engineering for animals is not here yet. And redesigning humans is even more distant. For with individual cells, a scientist with the right culture medium can select the one in a million that changed in the desired way, leaving the others to starve. But individual cells are not the same as individual human beings.

SOME IMPLICATIONS

We can think of many ways to use this technology, once it is perfected, but some of the applications will catch us by surprise. For these are powerful tools to help unlock the inner workings of the cell. With recombinant DNA, for example, the scientist no longer is stymied by the overwhelming genetic complexity of a cell. Now he can remove the DNA, chop it up into little pieces, put each piece in a plasmid, and multiply it in bacteria until he has enough of that piece to study in detail. We also may learn how cells lose control and become cancerous. For when cancer cells are injected into young mouse embryos, the chimeric mice that are produced have both genetic types of cells in their bodies, but they do not have cancer.[31] Somehow, the cancer was turned off. And this "switching off" does not happen only with cancer cells, for normal chimeric mice also are produced from embryos that receive cells from mice with a kind of muscular dystrophy.[32] Indeed, these genetic tools may even help us uncover the secrets of cell differentiation, the process of switching genes on and off that enables a single fertilized egg to produce the vast array of cells that makes up a person.

One of the most alluring uses of genetic engineering is to treat people with genetic diseases. The best candidates would be people with single gene defects. It may someday be possible to remove some of their cells, supply them in culture with the missing gene, select out the repaired cells, grow up a batch, and reimplant them in the patient.

Another prospect is to expand our repertoire for prenatal diagnosis following amniocentesis. Some disorders cannot be diagnosed because the fetal cells are not yet expressing the gene in question. Here the problem is to switch on the gene so it can be tested. Since fusion has this effect on some cells, it may be a solution. Another approach is to cut out specific genes from the fetal cells with a restriction enzyme and then chemically test them for abnormality. This remarkable approach recently was developed for sickle cell anemia.[33] A third possibility is to use these genetic techniques to map the location of genes on chromosomes. With a complete map of the human chromosomes, geneticists could diagnose disorders for which there is no direct test (yet), such as muscular dystrophy, by testing the genes that are close neighbors of the gene in question.

The most spectacular use of genetic engineering would be to redesign our present forms of life. Here our imaginations can run wild. A few ideas are: photosynthetic animals that use solar energy to make their own food; people with large brains and, presumably, great intelligence; advanced monkeys, or man-ape hybrids, that relieve us from tedious tasks; people who digest cellulose; people who live longer, perhaps forever; legless astronauts that fit inside space capsules; long-armed orchard workers; and tailor-made olympic champions.

Perhaps we also could make plants that produce more food at less cost. We could build in disease and insect resistance, and make strains with more high-quality protein. Scientists already have done this by radiation and selective breeding, but the new genetic tools will help them devise more precise and extreme combinations. They have, for example, produced hybrid plants by fusing cells from two types of tobacco.[34] Indeed, our rapid progress has prompted one scientist to remark: "Genetic engineering with plants is not science fiction any more. It's here."[35]

An intriguing idea is to eliminate our need for nitrogen fertilizers by producing plants that fix nitrogen from the air. Leguminous plants such as peas, soybeans, clover, and alfalfa have bacteria growing in their root nodules that convert nitrogen gas into a form of nitrogen the plants can use. But other plants lack these microbes, so when we grow those crops, we usually add nitrogen fertilizer, which is expensive to make (in terms of the raw materials and energy) and disruptive to our environment. We would solve those problems if scientists could either invent plants that have their own genes for fixing nitrogen, or design nitrogen-fixing bacteria that live in a wide variety of plants. A problem with the first idea is that plants produce oxygen, and oxygen blocks the enzyme for nitrogen fixation. If we could keep oxygen away from the bacteria, though, the second idea might work. Indeed, scientists already have spliced the bacterial genes for nitrogen fixation into a plasmid, fed the plasmid to another type of bacteria, and observed that the new strain fixed nitrogen when it was protected from oxygen.[36]

The list of possible benefits from reengineered microbes is truly awesome. For example, we might make microbes to clean up oil spills. In fact, scientists at General Electric have already produced a "superbug" to do just that. Another idea is to tailor bacteria to make protein from oil or, better yet, from such organic material as garbage and certain industrial wastes. That is well within our grasp. But scientists are scratching their heads over Greenland's plan to develop bacteria that can turn snow into oil.

One of the most exciting prospects is to use bacteria as miniature factories for special products we need. We could, for example, get microbes to produce large amounts of cellulase, an enzyme that converts cellulose into glucose. Then we could use that enzyme commercially to make glucose, a digestible food. Cellulose is the main ingredient in plant tissues and fibers and is the most abundant organic material in the world, so its conversion into glucose would dramatically increase our food supply.

Microbes could also produce rare and valuable substances for treating disease. If we gave them human genes, the bacteria would produce human-type proteins that would be ideal for human patients. With a good supply of specific enzymes, scientists could try enzyme therapy for people with leukemia or metabolic disorders such as PKU, galactosemia, and Tay-Sachs disease. And there would be a large market for human proteins that are hormones, such as growth hormone, ACTH (adrenocorticotrophic hormone), and insulin. An ample supply of clotting protein would greatly reduce the cost of treating people with hemophilia. Perhaps we also could design microbes to produce large amounts of antibiotics, interferon (a natural antiviral substance), and antibodies against particular diseases.

* * *

There is no doubt about it. Reengineering bacteria holds the promise of breathtaking benefits. And at least some of the promises will be kept. By the end of 1977 a research group in California had chemically synthesized a human gene, inserted it into bacteria, and obtained the protein it specified. The product was somatostatin, a hormone normally produced by the brain. From about four ounces of redesigned bacteria and two gallons of culture, the researchers recovered five milligrams of the hormone.[37] In contrast, the scientists who originally isolated somatostatin had needed nearly *half a million* sheep brains to get the same amount.

In 1978 a California group with many of the same members used recombinant DNA to help make human insulin. They synthesized two genes (one for each of the two protein chains in insulin), attached them to plasmids, and inserted them into bacteria. Then they isolated the individual chains from the microbes, purified them, and chemically joined them to form active insulin. According to one member, Arthur Rigg of the City of Hope National Medical Center at Duarte, "We're getting between 100,000 and a million molecules of insulin per bacterial cell."[38] Mass production appears to be in sight, and Eli Lilly and Company, an Indianapolis drug firm, is working on the project. A successful venture would greatly benefit the millions of diabetic people who rely on daily doses of insulin to stay alive. As things are now, it takes over four hundred pounds of pancreatic tissue (from cattle or pigs) to supply one ounce of insulin. With increasing numbers of diabetic people, increasing demand for insulin, the rising cost of insulin, and thousands of diabetics who are allergic to animal-derived insulin, human insulin by recombinant DNA would be spectacularly welcome.

SOME CONCERNS

Redesigning Microbes

With genetic engineering of microbes suddenly in our grasp, we find ourselves on the brink of some truly exciting benefits. But this also brings us face to face with an age-old dilemma: Scientific knowledge can be used many ways, only some of which are clearly desirable.

One concern is that scientists will accidentally produce an "Andromeda strain" that could escape from the laboratory and devastate the countryside. In recombinant DNA experiments the foreign genes are usually put in weakened strains of *Escherichia coli*, a bacterium that lives in human intestines. So some people fear the redesigned microbes—even those from weakened strains—might be able to pass their new tricks on to humans. Perhaps we would come to harbor genes for tumors, dangerous toxins, or strange new hormones. And we simply do not know enough about nature to predict what would happen in "shotgun" experiments, in which the entire DNA of an organism is cut into assorted fragments, spliced into plasmids that are inserted in bacteria, and then cloned (multiplied). Robert Sinsheimer, now chancellor of the University of California at Santa Cruz, has said: "Somehow it is presumed that we know, *a priori*, that none of these clones will be harmful to man or to our animals or to our crops or to other microbes—on which we unthinkingly rely. I don't know that and, worse, I don't know how anyone else does."[39]

The dangers are real, and the scientists themselves raised the issue. In 1974 a committee of the National Academy of Sciences, chaired by Paul Berg of Stanford University, published letters in *Nature* and *Science* asking scientists not to try the most dangerous types of recombinant DNA experiments until safety guidelines could be developed. Scientists throughout the world voluntarily complied. Seven months later an international group of 140 scientists met at the Asilomar conference center in Pacific Grove, California, and adopted provisional guidelines. The research resumed under those rules. In 1976 the National Institutes of Health (NIH) issued a similar set of safety rules, which were binding on all institutions receiving NIH grants.[40] Other funding agencies in the United States followed suit.

The NIH rules set up several classes of experiments. They ban experiments with tumor-causing viruses or genes responsible for dangerous toxins (diphtheria, snake venom, botulin). They also prohibit the release into the environment of any organism containing recombinant DNA. Other experiments are classified according to their potential for harm, and each carries a specific set of procedures for physical and biological containment. For studies requiring the highest level of physical containment (called P4), the materials stay inside sealed cabinets, and the scientists work by means of gloves that are attached to those cabinets. The complete facility also has a sealed environment, complete with elaborate decontamination systems. P3 facilities require open-fronted safety cabinets with a curtain of air across the opening, and negative pressure inside the laboratory to prevent air from escaping out the double-door entrance. P2 and P1 basically require standard, commonsense procedures for microbiological research. For certain, less dangerous experiments, researchers may transfer genes into the standard laboratory strain of *E. coli*, but they must use only plasmids that are not transmissible to other bacteria. The most dangerous work, however, is done with a mutant strain of *E. coli*, developed by Roy Curtiss III of the University of Alabama, that is coded to self-destruct if it escapes from the laboratory.

Is that good enough? Can we be sure there will not be a tragic accident? Since we live in the real world, we will not be able to get a 100-percent guarantee. Even if the rules themselves were impeccable, we could not erase the chance of human error, or expect perfect compliance. As things stand now, the NIH rules do not apply to private research, including industry, or to most work conducted outside the United States (though many countries have adopted rules of their own). And even where these rules do apply, there are practical problems in implementing them.

The rules are not perfect, either; they are based largely on educated guesses about hypothetical dangers. Nevertheless, if they were followed by all research workers, they certainly would minimize the risk of an accidental epidemic. Indeed, the containment procedures (P4) parallel those used for research in biological warfare, which involved the most virulent organisms known to man. At Fort Detrick, where germ warfare was researched for twenty-five years, only one worker who became infected transmitted the disease to someone else.[41] The experiences of the Center for Disease Control (United States Public Health Service) and of laboratories in other countries also attest to the high level of safety with those facilities. The risk of an epidemic from gene-splicing research is small in comparison, for the NIH rules ban experiments with such dangerous microbes. Indeed, we are more likely to be endangered simply by our human error and ignorance. According to Sydney Brenner of Cambridge University, who helped frame the agreement at Asilomar: "The greatest biohazard is our lack of knowledge. The fact is we don't know. And what we don't know, we can worry about."[42]

While scientists and others have engaged in a sometimes shrill debate over the risks of an accident, another danger has received precious little attention: engineering an "Andromeda strain" on purpose.

The scenario might go something like this: First, Dr. X inserts the genes for several deadly diseases and toxins into a bacterium that can infect man. Then he stirs in some genes that make Wonder Bug impervious to our usual medicines. Now he has created a weapon that only one or two obscure antibiotics can stop. Dr. X's group secretly stockpiles those antibiotics to protect its members, leaving everyone else at their mercy. Then he issues his ultimatum.

Impossible? A microbe might not accept such virulent genes and survive. And there would be severe tactical problems. Furthermore, many nations, including the United States and the Soviet Union, have signed the Biological Weapons Convention, thus declaring that they will not use germ warfare. Still, some nation or terrorist group might find the idea attractive. The superbug, like the neutron bomb, would cause little property damage. And the cost might be low enough that each group could develop its own, at least for the sake of self-defense. In time, perhaps, no self-respecting nation would be caught dead without one.

Treating human disorders

One dream of the genetic engineers is to repair hereditary defects in people. At the moment they have two possible approaches. One is to infect the patients with viruses that carry the right DNA. The other is to remove some of their cells, repair them in culture, grow up a large batch, and reimplant them.

But it is not as easy as it sounds. One problem with the virus approach is finding a safe virus. We know of viruses that insert DNA into human cells, but they also may cause tumors. And viruses usually will not carry the desired gene(s), so scientists will have to splice the right DNA on the virus and hope it still can infect human cells safely. Another problem is getting the virus to infect enough of the right cells while being attacked by the body's immune system. Yet even if we could overcome all those obstacles, we might find that the reengineered cells looked like strangers to the immune system and were destroyed.

The virus method is not exactly trouble-free, so let us consider the other idea. Removing some deficient cells and growing them in culture would not be too hard. But how would we redesign them? Using viruses would raise the same problems we just discussed, except there would not be a rejection problem in the culture medium itself. Plasmids might pose fewer problems, but as far as we know, normal human cells do not contain them. Whole chromosomes would rarely produce the desired results. Reverse mutations would be a very long shot. And fusion is too imprecise. And whatever method we tried, we would soon confront another danger: After a while in culture, cells tend to change into cancer cells. Normal cells from adult humans survive for only about twenty generations in culture,[43] while many tumor cells reproduce indefinitely. So by the time our genetic engineer removed the deficient cells, cultured them, repaired some of them, selected out the winners, and grew up enough for implantation, the patient might be better off to take a rain check.

Two other problems loom over all these methods. Unless we could supply only the desired genes, the redesigned cells might have some extra, unwanted genes. And Down's syndrome is a sober reminder that too much genetic material can itself cause abnormalities. The other problem is controlling how the cell would use its new gene(s). If the gene were not expressed in the cell, the therapy would fail. Yet if the gene were used but were not regulated properly, the cure could be worse than the disease. Suppose, for example, that a diabetic person received the genes for insulin, but his pancreatic cells could not control them. If he produced too much insulin, his blood sugar level would plunge, producing a coma that could be fatal.

So we can see that gene therapy is a risky proposition, at least for the immediate future. Are there, then, any situations where we might consider it worth trying? Theodore Friedmann of the University of California, San Diego, and Richard Roblin of the Harvard Medical School have proposed the following criteria: The disorder is understood well enough to establish that gene therapy might help; there is prior experience with untreated cases, so scientists can assess how well the therapy works; the DNA supplied is sufficiently pure; the method has been tested thoroughly in experiments with animals; and where appropriate, the treatment is first tried on cultured skin cells from the patient.[44]

These are worthy criteria, but they will not always be met. Animal testing will be incomplete in many cases, for animals do not have all the same genetic diseases as people. And there will be occasions where desperate patients want to take a chance on a treatment that is untested and dangerous but offers a glimmer of hope.

Are we ready to use gene therapy on humans? No, but it has already been tried. Two young sisters had hyperarginemia, a hereditary disorder where the enzyme to metabolize arginine (an amino acid) is absent. As a result, toxic levels of arginine accumulated in their blood, causing them to be severely retarded. There is no effective treatment for this disorder, and their condition was considered hopeless. As a last resort, the girls were injected with an apparently harmless virus. The virus was thought to carry a gene for their missing enzyme, and it had previously lowered the arginine levels in the blood of laboratory workers who had been infected accidentally. The virus temporarily reduced the girls' arginine levels, but it had no effect on their retardation.[45] So this first experiment with gene therapy proved neither helpful nor harmful. That it was done at all, though, suggests we will see other attempts in the future, especially where there is little to lose.

Yet even if we can make gene therapy safe and effective, it will be expensive in terms of cost versus benefit. It could help people with single gene defects, but it would not be of much use to those who have dominant, polygenic, or gross chromosome disorders. There are hundreds of different defects that might respond to gene therapy, and each would likely require a different piece of DNA, but collectively they occur only an average of once every one thousand births.[46] According to Walter Bodmer of Oxford University and Alun Jones, a British scientist and writer: "The problem with gene therapy is that it will be a sophisticated method of curing very rare diseases. The result is that when a cure for a particular disease is developed, the cost of each successful diagnosis will be high. Should the money now available therefore be spent on developing the best form of treatment for sufferers of these diseases, or on supporting research based on gene therapy?"[47]

So genetic engineering is at best a faraway solution for individual tragedies. And if it ever became feasible, it could provide further deterioration of our gene pool. For if we ever reached the point of repairing body cells but not egg and sperm cells, many "cured" patients would survive to pass their defects on to their descendants, generating still more customers for gene therapy. Then when we decided it was time to sanitize our gene pool, we would have to require all our fertile citizens to come in for repairs. But since we cannot prevent spontaneous mutations, we would only be able to issue sixty-day warranties on their genetic fitness. Indeed, they would need at least semiannual tuneups.

Redesigning Man

Behind the clamor over genetic engineering lurks the ultimate concern: redesigning people. That prospect lies far on the horizon, to be sure, but few scientists would dare predict that we will never have that power. So maybe it is worth spending a few minutes in dreamland.

One problem would be to decide which features we wanted to build into man. Some of the qualities we might like—body size, artistic and mathematic aptitude, intelligence—depend on several different genes, plus powerful environmental factors. So they would be especially hard to incorporate. We also might discover that some of the features we chose were incompatible. For example, Joshua Lederberg has speculated that it may be impossible for people to have a minimum risk both of cancer (which may reflect a low level of immune response) and allergies (which result from an active immune response).[48] And we might find there is a genetic link between two traits, only one of which we want. Perhaps musical ability is somehow related to the temperament (un)popularly associated with such artists. We know so little about the genetic basis of behavior that this may well be a poor example. But of course that is a major problem in trying to redesign man: We know so little.

And who would choose the "superior" features? And how would they implement the program? According to C. P. Snow: "I can't think of any individual people who could be trusted with such a power. Or any society. The power to produce what they thought of as the most estimable or valuable or useful human beings: in a primitive pre-scientific fashion, the Nazis tried this by selective breeding for the perfect Nordic man. Our choice of ideal human beings might be more tolerable, but not much."[49]

Now let us look at the other side of the coin. First of all, we can dismiss the objection that genetic engineering would necessarily be wrong because it would be meddling with nature. The issue is *not* whether we may "meddle," but whether the benefits of doing so outweigh the costs. At first we would use genetic engineering only to relieve the suffering of people with genetic diseases. Indeed, we could hardly justify withholding this technology from them just because we were afraid of misusing it later. Then after we learned how to improve them genetically, we would gradually start helping other, more "normal" people enjoy the same advantages as those who fared better in the genetic lottery. We would have to be careful not to use genetic engineering so much that it would undermine our genetic diversity. But since we realize how important diversity is, we probably would provide for it in our genetic plan.

Genetic engineering could be our time machine to the future, speeding us toward perfection. Throughout our long history, our genetic improvement has been controlled by the slow, mindless process of evolution. But now we see a way to break away from that pattern, installing ourselves as the masters of our genetic future. The prospect is breathtaking. Indeed, the late Pierre Teilhard de Chardin wrote: "The dream which human research obscurely fosters is fundamentally that of mastering, beyond all atomic or molecular affinities, the ultimate energy of which all other energies are merely servants; and thus grasping the very mainspring of evolution, seizing the tiller of the world."[50]

Should we seek that power? On the one hand, we could argue that we are in no great need of genetic improvement, except for correcting defects. The potential benefits simply would not outweigh the risk of a tragic mistake. Furthermore, our values as to what genetic features are best would change as we changed, so we would be setting out on a genetic odyssey with no known destination.

On the other hand, we rarely can foresee all the consequences of our actions; redesigning man would be no different. And history is littered with misguided people who opposed scientific "progress." We seem not only to have survived those advances but to have benefited from them. So once again, perhaps, we would need the courage to continue our ascent. Yet the stakes are higher than ever before. As the science writer Nicholas Wade has explained: "Other technologies developed in the course of civilization are merely extensions of man's hands or senses. The ability to manipulate the stuff of life is an art of a different order, the ultimate technology."[51]

That is enough mental Ping-Pong for now. We do not have to settle the issue this week, so let us ring down the curtain on this fantasy with a pair of contrasting views. Leon Kass offers a chilling appraisal: "By seeking to transcend our nature, we may fall far below it. By seeking to redirect evolution, we may spoil the fruit of previous evolution. Creatures of human shape might survive, but human beings might become extinct."[52]

In his poignant essay "The Brain of Pooh" Robert Sinsheimer expresses a bit more optimism: "Perhaps when we've mutated the genes and integrated the neurons and refined the biochemistry, our descendants will come to see us as we see Pooh: frail and slow in logic, weak in memory and pale in abstraction, but usually warmhearted, generally compassionate, and on occasion possessed of innate common sense and uncommon perception."[53]

Death is an imposition on the human race, and no longer acceptable.

The Immortalist by Alan Harrington

The wise man lives as long as he should, not as long as he can.

Seneca

7

The Fountain of Youth

We know what lies in store for us: gray hair, stiff joints, fragile bones, wrinkled skin, hearing aids, thick glasses, slow reflexes, poor coordination, weakness, lapses of memory, greater risks of infections, heart attacks, arthritis, cancer, and diabetes. It is not pretty. We would all like to escape those effects of old age, but we also would like to live a long time.

How long would you like to live—seventy, one hundred, three hundred years? Forever? Our answers depend largely on the quality of those extra years. Living them at the functioning level of a seventy-five-year-old is not so appealing, but if we could continue to enjoy the vigor of the early and middle years, most of us would jump at the chance. Ernst Wynder, president of the American Health Foundation, put it another way: "It should be the function of medicine to have people die young as late as possible."[1]

What can we expect? Is the fountain of youth only a mirage, or will we learn how to stem the ravages of time on our bodies? The answer lies deep within the heart of an ancient mystery—the aging process. We must begin there.

MYSTERIES OF AGING

People have been getting old for a long time, but we still do not know why or how it happens. One reason is that it is a difficult subject to study. For one thing, the effects of aging are so diverse that there may well be several different causes. And even the effects are hard to identify, for they vary with each person, and they are entangled with the effects of diseases. People generally do not die from "old age," but from specific disorders. Yet many degenerative diseases seem to go hand in hand with old age.

Scientists can try to sort out those effects by studying large numbers of people and doing statistical analyses (also known as the art of data massage). The best data would come from testing the same group for many years. But

87

that is easier said than done. The subjects would have to stay accessible and cooperative for several decades. And the researchers would have to forego publication for a while and the new grants those publications might have elicited. They would also need to stay alive.

Since scientists literally get gray hair doing such studies, many work instead with animals that have short life spans. Rotifers are a popular choice because they live only a few weeks, but they would seem a poor model for human aging. Mice and rats are a better bet, but they live about three years. That is a long time for an experiment (though it is just about right for students who need a doctoral thesis project). Another time saver is to accelerate aging by exposing the animals to radiation or high temperatures. Yet we can hardly be confident that this is simply a fast-motion version of normal aging. For example, a careful study of the people in Hiroshima revealed no correlation between their rate of aging and their exposure to radiation from the atomic bomb.[2]

Despite all the difficulties, or because of them, there are many hypotheses about why we age. One of the major themes is that we gradually accumulate errors in our bodies and do not repair them. Our body parts do not deteriorate at the same rate, and the most vulnerable tissues seem to be those that can not readily replace damaged cells with healthy ones. The prime candidates are cells that do not divide at all (muscle and nerve), or do so rarely (kidney). Cells can often repair themselves, but in time the repair machinery may itself need repair; then the errors accumulate faster. As individual cells stop functioning, the total capacity of the heart, other muscles, kidneys, and brain goes downhill.

The error theory fits with many of the symptoms of aging. According to a long-term study of seven hundred normal men that began in 1960, the major changes include: a lower capacity for consuming oxygen during strenuous exercise, less muscular strength, slower transmission of nerve impulses, poor coordination, a declining ability to reason (especially after age seventy), and a longer time for kidneys to correct chemical imbalances in the blood.[3] Other studies have suggested that older people have fewer functioning cells in certain vital tissues. For example, a widely quoted (but unsubstantiated) estimate is that we lose one hundred thousand nerve cells a day during old age. That figure probably is much too high, but there is little doubt the loss is considerable. Bernard Strehler, a gerontologist at the University of Southern California, has estimated that after the age of thirty, our reserve capacity for a wide variety of functions declines by nearly 1 percent per year.[4]

The error theory "explains" many effects of aging, but it is too vague to help us pinpoint the actual cause(s). We need to ask specifically what those errors are, and what causes them. Here there is no shortage of candidates, so let us briefly survey seven leading contenders.

Contestant number one is the so-called clinker hypothesis. As we get older, brown granules accumulate inside certain of our cells, occupying as much as 25 to 30 percent of the volume of nerve and heart cells in old people. This "age pigment" comes from the incomplete breakdown of fats in the cell. According to the clinker hypothesis, cells eventually become so clogged with this material that they can not function properly.[5] Yet the embarrassing fact is that there is no evidence this material actually does interfere with cell activities. Indeed, "age pigment" is probably an effect of aging rather than a cause; it is basically a problem of garbage disposal. Thus the term "clinker hypothesis" seems doubly appropriate.

The second candidate, one that overlaps several others, is the idea that we age because increasing numbers of our body cells are damaged by spontaneous mutations. If not repaired, many of those changes could be harmful, especially in cells that do not divide. Mutations could also damage the repair systems. That all seems reasonable enough, but the direct evidence for this hypothesis is a bit underwhelming. It does fit with the dubious claim that radiation accelerates aging. And there is an age-related increase in chromosome abnormalities in rat liver cells, but not white blood cells (which are rapidly replaced).[6] But while there is little doubt we carry an increasing burden of mutations as we grow older (some come from the radioactive elements in our own bodies and cosmic rays), that does not prove mutations actually cause aging.

A related idea is that we age because our cells gradually fail to make the right proteins. This could happen because of mutations, but it also could arise from errors in making RNA from DNA, or in using RNA to make proteins. Since proteins themselves participate in the making of proteins, and in the repair of DNA, errors in those particular proteins could set off a cascade of other errors. The cell would produce increasing numbers of abnormal proteins, finally losing its ability to function.[7] It is an intriguing idea, but there is a practical problem: Because of the chicken-and-egg relationship between DNA and protein, scientists find it very difficult, experimentally, to separate this hypothesis from the preceding one and test it.

According to contestant number four, as we age, our proteins and other complex molecules unravel, gradually losing their ability to function. Materials that are not readily replaced would be especially vulnerable. This idea fits with the fact that enzymes and other proteins need a precise, three-dimensional structure in order to work properly. But we are still waiting for evidence that this loss of structure is a major event in cells, and that it is a cause of aging.

Our next candidate, the cross-linking hypothesis, asserts that aging occurs because large molecules react (cross-link) with each other to form non-functional blobs. Although cross-linking can involve DNA, most of this research has focused on proteins. One version is that inactive protein complexes clog the cells and eventually destroy them. It is the clinker idea again. Another, more respectable version pinpoints the structural proteins, collagen and elastin. Muscle, skin, bone, cartilage, blood vessels, and other structures all have large amounts of these proteins, so their cross-linking could have far-reaching effects. Indeed, cross-linking does cause collagen to shrink and become more rigid, so this could contribute to lower heart and lung capacities, stiff joints, and wrinkled skin. Cross-linking could also upset the normal exchange of nutrients and wastes between cells and the blood. But whereas these features of the hypothesis are impressive, there is a serious question as to whether cross-linking is a cause or an effect of aging. For one thing, there is no evidence that age-associated cross-linking occurs in cells.[8] Another objection is that collagen cross-linking, which occurs outside cells, is largely complete by age forty.[9]

Proposal number six singles out our immune system as the villain.[10] As we grow older, our immune system becomes less effective in warding off invaders, leaving us more vulnerable to infections. We also become more susceptible to autoimmune diseases (rheumatoid arthritis is a likely example), where our own antibodies attack ourselves. Yet these problems must stem from some underlying error, so they may be an effect of aging rather than a cause. One possibility is that antibody-producing cells eventually become damaged so they cannot produce the usual amount of antibodies. There also may be a breakdown in the way our immune system distinguishes "self" from "foreign." Or perhaps spontaneous mutations, or other errors, change our cells into "foreigners," thus inviting our immune system to attack them.

The seventh hypothesis, which encompasses many of the others, is that free radicals cause aging. Free radicals are molecular fragments that have an odd, unbalanced number of electrons. This makes them very reactive. They are produced naturally in the body, especially by reactions involving oxygen. Like their political namesakes, free radicals produce an unstable environment where changes rapidly occur. Because of their lust for electrons, they indiscriminately attack other materials, thus upsetting the delicate chemical balance in an organism. Their effects include cross-linking, genetic damage, the disruption of cell membranes, and the breakdown of fats into "age pigment."

The free-radical hypothesis looks good on paper, but there is a shortage of convincing evidence. The main support comes from studies with antioxidants, chemicals that react quickly with free radicals to quench their thirst for electrons. Antioxidants in the diet have increased the average life span of mice by as much as 30 to 40 percent.[11] And vitamin E, a natural antioxidant, increases the life span of flies by 13 percent, prevents capillary damage in chicks, and helps protect rats against free-radical damage to their lungs.[12] But whereas antioxidants may increase the average life span, it is not clear that they increase the maximum life span. And the distinction is important; any drug (for example, penicillin) that protects against premature death will increase the average life span, but this does not necessarily have anything to do with aging itself. Furthermore, there is no clear evidence that humans (unlike many animals) need vitamin E to be healthy. And the credibility of vitamin E as a curative agent is not helped by claims of its effectiveness against rough skin, atherosclerosis, declining virility, muscular dystrophy, diabetes, ulcers, inflation, and violence on television.

All the ideas we have examined so far fit the basic theme that aging is caused by a gradual accumulation of errors; although individual errors may occur randomly, the total wear and tear of life will catch up with us at a fairly predictable age unless something else kills us sooner. Indeed, our experience suggests man may have a maximum life span of about 120 years. Few people have lived longer than that, and even fewer can document their claims. But some scientists interpret the idea of an upper age limit in a different light: They believe we are internally programmed to self-destruct.

That brings us to one of the oldest contestants—the rate-of-living hypothesis. According to this idea, we have a specific, maximum amount of energy to spend in our lifetime; when it is gone, so are we. Among different types of animals there is in fact a striking correlation between their metabolic rates and their maximum life spans. For example, man lives about thirty times longer than the mouse, and his metabolic rate is corresponding slower. Furthermore, factors that lower the metabolic rate increase the life span. Animals that hibernate, or naturally have lower body temperatures, have longer maximum life spans.[13] And young rats on very low calorie diets mature slower and live longer than rats on normal diets.[14] Underfeeding also boosts the life spans of rotifers, silkworms, fruit flies, bees, chickens, and other animals.[15]

But despite these impressive correlations, the rate-of-living hypothesis is too simplistic for us to accept at face value. Even if it were correct, it would not really be an answer; it would still leave us with the crucial, unanswered question of what causes an energy limit to exist. (Nevertheless, this hypothesis does have one outstanding virtue—it gives us a powerful excuse for being lazy.)

If we are programmed to self-destruct, we must have a clock embedded in us somewhere. But where could it be? Some researchers believe it may be in our genes. They can point, for example, to experiments where the same patch of mouse skin was periodically retransplanted onto young mice of the same genetic type; despite always being on a young mouse, the skin always died. Sometimes the skin lived twice as long as a normal mouse, but that discrepancy is not surprising because skin probably is not the limiting factor in how long a mouse lives. The problem with this experiment, however, is that each operation damaged the skin and disrupted its blood supply. Thus the dear, departed skin might have the transplants to blame rather than its own clock.[16]

Leonard Hayflick, now at Stanford University, made the startling discovery in 1965 that normal cells have a finite lifetime in culture.[17] Human embryonic cells multiply about fifty times in culture before dying, and the limit is about twenty for adult cells. A few substances can increase the limits slightly, but freezing has no effect.[18] And as the cells age, they seem to slow down and take longer to divide. Although these results suggest cells have a built-in clock, we must be careful how we interpret them. The basic problem is that the aging of cells in culture is not necessarily the same as the aging process in the body. For one thing, cells grow much faster in culture than in the body; in fact, in the body they probably do not reach the limit observed in culture. Indeed, old and young adult cells double about the same number of times in culture.[19] Furthermore, many of our normal activities, and many effects of aging, reflect the complex interplay between our cells, tissues, and organs. Those interactions are absent in culture.

Some scientists think the brain may harbor a death clock. A specific candidate is the hypothalamus, the part of our brain that is the master switchboard for our hormone secretions. Since hormones help regulate our metabolic rate, and they trigger changes at fairly specific times in life (early development, puberty, and menopause), it is logical to think they also might time life itself. Yet the evidence for this idea—like all the others—is meager, and the nature of the clock (if there is one) remains a mystery.

This is about as far as we can go in our brief survey. We find that the secrets of aging are not in any immediate danger of being discovered. Indeed, some of the theories on aging are themselves getting old.

What, then, can we expect in our quest to slow down or prevent aging? The prophets seem to be at a loss on this one, for we can find virtually any forecast we want. Robert Preholda, who wrote a book on aging, predicted for 1981–1990: "Antiaging pills that cost no more that 30¢ a day will be available. Composed of antioxidants, such as vitamin E, these will neutralize free radical molecules . . . The effect on society and human life will be tremendous."[20] And Alex Comfort, formerly the director of gerontology research at London's University College (who is perhaps better known in most circles as the author of *The Joy of Sex*), contends that by 1990 chemicals or special diets will increase our life span by 20 percent.[21] Another recipe, offered by Bernard Strehler, comes from the observation that lower body temperatures mean longer life spans for many animals. He said: "If this rule applies to humans, as it does to all other systems studied so far, reduction of body temperature to 35°C [from 37°C] would add about 15 to 20 years to healthful life."[22]

A modest increase in average life span is one thing; immortality is quite another. Even those who are optimistic about the first prospect are less sanguine about the second. According to Comfort, there is "no reasonable prospect of securing effective personal immortality for anybody in the foreseeable future."[23] Strehler, on the other hand, declared: "Someday we may live almost indefinitely."[24]

Several octaves lower on the euphoria scale is John Maynard Smith, who wrote:

It is not at present possible to say whether we shall ever be able to produce a large increase in human life expectancy, even though we can already ensure that a larger proportion of people survive to old age. We do not at present know whether senescence is caused by a number of physiologically independent processes—in which case, even if we prevented one of these processes people would still die at much the same age of another—or whether there is one fundamental process of which the various superficial signs of senescence are merely symptoms. If the former assumption is correct, and the evidence suggests to me that it is, then a significant extension of the human life span is likely to prove very difficult.[25]

Smith expressed that view in 1965, and it still stands. In the dim light of our subsequent progress in understanding aging, forecasts of dramatic and imminent increases in our life span amount to whistling in the dark.

REPLACING BODY PARTS

We may not halt the aging process itself, but we are learning how to help more people reach old age. And by replacing parts of our bodies that are wearing out because of disease or age, we may escape, at least temporarily, some of the infirmities of old age. Now let us examine three ways of doing this.

Transplantation

For many centuries the idea of using body parts from another person or animal was only an interesting idea. Not until the nineteenth century did physicians realize they could transfer pieces of skin from one part of a person's body to another. Transplants (of cornea and bone) between people began at the turn of this century. The first successful kidney transplant in humans dates back only to 1954, and within the next fifteen years surgeons tried heart, liver, lung, thymus, bone marrow, spleen, intestine, and pancreas transplants. Now they are working on muscle, limb, adrenal gland, thyroid gland, testicle, ovary, oviduct, pituitary gland, and other transplants. The pace is breathtaking.

Of the major organs, kidney transplants have been the most common and successful. In the United States alone, more that three thousand people a year receive kidney transplants. And people already have lived more than a decade with their new kidneys.[26] About half the kidneys transplanted from cadavers have functioned well for at least two years, while the success rate is 70 percent or better with live donors who are blood relatives of the recipients.[27] And the patients fare even better than their new kidneys; for if their first transplant fails, they can use "artificial kidney" dialysis machines while awaiting another try.

The record is not so impressive with other major organs. Lung and pancreas transplants have not worked well. After a decade of poor results, however, the success rate for liver transplants has improved; now the one-year survival rate is 40 percent.[28] Heart transplants generated tremendous interest when South Africa's Christiaan Barnard performed the first such operation on a human in December 1967. His patient, Louis Washkansky, died eighteen days later. Although the enthusiasm for heart transplants has dimmed considerably since then, there has been steady improvement in the results: The one-year survival rate has increased to nearly 70 percent; 25 percent now survive for five years; and the longest survival period, as of early 1979, was about nine years.[29]

We can expect success rates to climb further, especially as we develop better weapons against rejection. But that will not be easy. For with a few exceptions (especially the cornea), our body parts come under the protective eye of our immune system, which regards all strangers, even life-saving organs, as enemies to be destroyed.

One tactic in transplant operations is to use drugs that suppress the immune system. It is a delicate balancing act. If the dose is too small, the transplant is usually rejected. This can be fatal with heart, lung, liver, and other transplants where there is not an artificial organ to rescue the patient. Yet too much drug also can be fatal, for it leaves the patient helpless against infections and cancer. Furthermore, the drugs themselves are toxic. One physician observed that "the most serious complication of steroid therapy is death."[30]

Another tactic is to tone down the rejection response by matching the donor with the recipient. Tissue can be "typed" according to its antigens (substances that provoke the body to produce specific antibodies). If the recipient's own cells carry the same antigens as the transplanted tissue, his immune system will find nothing to reject. Unfortunately, the number of antigens is so large that only identical twins are likely to be a perfect match.

Nevertheless, a computer can compare the tissue types of donors with those of potential recipients to find the best matches. The larger the supply of organs, the better the odds will be that a needy person will receive a well-matched organ.

But that brings us to a sticky problem—supply. There are only three possible sources: living persons, cadavers, and animals. Live donors can supply only nonvital tissues, including the spleen, bone marrow, single members of organs that come in pairs (kidneys, adrenal glands, parathyroid glands, and gonads), and portions of skin, small intestine, and thyroid tissue.[31] So for many human tissues, cadavers are the only source.

The problem with cadaver organs is that they rapidly deteriorate once they stop receiving oxygenated blood. Kidneys, for example, should be removed from the body within half an hour of death if they are to be transplanted. It would be possible to preserve organs inside dead bodies, particularly in cases of brain death, as long as machines kept oxygenated blood coursing through the cadaver. But there obviously are limits on how long this would be practical and desirable. A better solution would be to remove the organs at death and preserve them until they were needed. Indeed, scientists are studying the effects of low temperatures (which slow degenerative changes), high pressures (which force oxygen into tissues), and perfusion with fluids that carry oxygen, nutrients, and drugs. Yet even under the best conditions found so far, they have only been able to preserve isolated organs for a few days. The current record—with rat hearts—is nine days.[32]

The best method for long-term storage would be to freeze the tissues. But it is not quite that simple. One problem is that water expands when it freezes, so ice would rupture cells and other structures. And ice leaves behind high concentrations of salt, which can also be harmful. Scientists have discovered, however, that there is less damage when they first replace some of the water with protective agents such as ethylene glycol, the main ingredient in antifreeze. They also are working out the optimal rates of freezing and thawing, and the best storage temperatures. Already, they have succeeded in freezing, storing, and reviving skin, bone marrow, corneas, sperm, young embryos, and blood cells. But freezing complex organs is far more difficult. For one thing, organs consist of several types of cells, which are likely to differ in their optimal conditions for freezing, storage, and thawing. And organs are thick enough to provide insulation, so the cells, even of one particular type, cannot all cool and thaw at the same rate.

Nevertheless, progress is being made. G. L. Rapatz of the Cryobiology Research Institute at Madison, Wisconsin, managed to freeze frog hearts for a few minutes and revive them enough to resume their rhythmic pattern of contractions.[33] Scientists also achieved a very limited freezing and revival of rat hearts.[34] Still, that is a far cry from freezing complex human organs and storing them for transplantation. Because of their size, however, embryonic organs might be better candidates. Indeed, scientists were able to freeze fetal rat pancreases, store them for forty-eight hours to thirteen weeks, transplant them into rats, and have about half of them function reasonably well until they were rejected.[35]

Freezing human organs in a viable state will be difficult enough, but the odds skyrocket for doing this with whole bodies. This would not be an efficient way to store organs for transplantation anyway, but some people see another use for this technology—a sort of immortality. For example, people with fatal ailments could cool their heels until scientists discovered a cure for their disease. Or at least they could preserve the chance—assuming that human cloning comes to pass—of eventually bestowing on their families, and society in general, the ineffable benefits of having their genetic copies around indefinitely.

I have not noticed any angels treading in that direction, but a few people are. By 1976 at least twenty-four human bodies had been frozen and stored.[36] According to one recipe, the first step is to flush a protective fluid through the cadaver by means of the carotid artery and the jugular vein. Next, the body is wrapped in plastic and cooled on crushed ice and salt until it is below the freezing temperature. Then it is wrapped in aluminum foil and stored on dry ice. Finally, the cadaver enters its permanent storage facility, essentially a large thermos bottle chilled with liquid nitrogen.[37] Although there have been a few takers, the public attitude has remained cool, and with good reason. In view of the technical problems, the minimal prospects for revival, and costs in excess of fifty thousand dollars for preparation and indefinite storage, the clients seems to be getting nothing for something.

Another way to increase the supply of organs is to make greater use of animal "donors." But one of the drawbacks is that animal organs are more likely to be rejected in humans. Furthermore, organs have evolved to operate efficiently in the chemical and physical environment of their host, so what works well inside an aardvark will not necessarily work inside a human. Yet animal organs also have certain advantages over human organs: There are fewer ethical and legal problems in removing the organs; we could better control the supply of organs to meet the demand; animals could be selectively bred to minimize rejection problems; and it may be possible to induce tolerance in the recipients. Most complex organs would come from chimpanzees, which most resemble man but are fairly rare, or from their more abundant cousins, the baboons. Another candidate, whose chemistry resembles that of man, is the pig. Perhaps there is a moral in there somewhere.

Our experience with transplanted animal organs is not too encouraging. The first attempt to implant a chimpanzee heart in a person occurred in 1964. The heart was too small, and the patient died ninety minutes after the operation. Transplants of chimpanzee kidneys have also failed, though one patient survived for nine months.[38] On the other hand, heart valves from pigs have worked well in humans, probably because the valves, like corneas, are not so vulnerable to rejection.

In most cases people who need transplants are better off with human organs. But we can expect a wider use of animal organs as a last resort, or as a way to buy time for the patient. For example, Christiaan Barnard has developed a technique for implanting an animal (baboon or, preferably, chimpanzee) heart to assist the patient's own heart until it recovers.[39] And in 1968 his medical group reported yet another way to use animal organs. Their patient, a woman dying of liver failure, could not remove certain toxic materials that were accumulating in her blood. In a final attempt to save her life, they cross-circulated her blood for six hours with an anesthetized baboon that had been transfused with a compatible type of blood. Three days later, the woman made a dramatic recovery.[40] The report did not mention how the baboon fared.

Regeneration

Another way to replace body parts is to regenerate them. When hydra and flatworms are cut in half, each segment regenerates a whole animal. Lobsters, crabs, cockroaches, grasshoppers, caterpillars, crayfish, shrimp, newts, and toads all can replace legs, claws, or antennae. Certain fish regenerate finlike structures, and some can even replace nonvital parts of the brain. Many vertebrates, including fish and birds, regenerate the optic nerve. And even mammals can play the game; for example, deer regularly grow new antlers and sheep grow new horns.

What happened to us? Why are we left out? Did we lose this power somewhere along our evolutionary path, or is it lurking inside us, waiting to be activated? We do not know.

We do know, however, that our bodies can partially compensate for certain losses. When we lose one of a pair of organs, or part of a single organ, the remaining tissue often enlarges. This happens, for example, with the pancreas, liver, thyroid gland, kidney, and adrenal gland. One of the most dramatic examples of mammalian regeneration is the mouse liver, which can be reduced to as little as 30 percent of its normal size and regain its normal size in a week. Depending on the organ, and the ability of its cells to multiply, this compensatory growth occurs by enlarging existing cells, or increasing the number of cells, or both.

Despite considerable study, we know surprisingly little about how regeneration works. We know a regenerated part does not always look just like the original, but it does restore the original function. And we know the process involves neural and hormonal signals, and unspecialized cells, but the details are murky. Another complication is that regenerating a body part is not necessarily the same process as growing the original part. There are several examples where embryos fail to regenerate parts that the adult animals can replace.[41] So apparently regeneration is a specific ability an adult may develop; it is not just a localized rerun of embryonic development.

Until we know more about regeneration, we can only guess whether someday we will replace our body parts this way. Scientists are trying to stimulate regeneration by manipulating the electrical signals in the region near the lost part. And they are learning more about how severed nerves can grow back in place. But despite all their efforts, nature suggests we may never have the powers of regeneration we would like. For the most complex animals generally have the poorest ability to regenerate, and that probably is not a coincidence. Since complex parts require a high degree of cell specialization, and regeneration requires less specialized cells, regeneration may be a casualty of specialization. Although that would dim our chances of regrowing body parts, at least we could drown our sorrows with the following rationale from Richard Goss, a biologist at Brown University:

Man is in no position to dispute the dogma that what evolves is by definition that which is fittest. Our very existence on earth testifies to the fitness of our progenitors, and hopefully of ourselves. This, despite the inability of most of us to regenerate parts of ourselves we might often wish to replace. Yet we have eluded extinction despite such shortcomings. . . . Thus, if we lack certain attributes that other animals possess, let us not lament the unfairness of things, for what we do not have would probably have cost too much anyway.[42]

But even if we could not regenerate parts naturally, we still might have another way—growing them outside our bodies and then implanting them. James Bonner, a biologist at the California Institute of Technology, has speculated: "Maybe you will go to the doctor and he will say, 'Well I think maybe your heart isn't so good now. Maybe we had better start growing you a new one. In another 2 or 3 years, it will be grown up, and we can plumb it in.' "[43]

It will take some doing. Since we probably would need to start with fairly unspecialized cells, preferably of the same genetic type as the future recipient, the best bet would be to collect some of the recipient's own cells prior to, or at, birth and store them for several decades until he needed them. Then we would grow the cells in culture. We would have to convince them not only to make copies of themselves, but also to develop into the right combination of specialized types of cells, and then to organize themselves into an organ. And we would want them to do this fast enough that the patient would not have too long a wait in the doctor's office. But we would also have to keep them from growing so fast that they would die, or become cancerous, before the organ was finished. Overall, then, the idea of growing replacement parts in culture sounds good, but there is a bit of a chasm between theory and practice.

Another possible way to grow replacement organs is to use animals as incubators. In one series of experiments, human fetal tissues (lung, thymus, testis, ovary, pancreas, adrenal gland, kidney) grew for two months in the abdominal walls of mice that lacked an immune system.[44] Several tissues seemed to develop normally there. This, of course, is not the final answer; those mice have a maximum life span of only five months, and they are a bit small for growing most human organs. Even more important, we do not know whether normal human organs can fully develop inside any kind of animal. Nevertheless, this experiment may give us a glimpse of things to come.

Artificial Body Parts

We will increasingly be able to trade in our broken-down body parts for mechanical devices. According to Vincent Gott of the Johns Hopkins School of Medicine: "Eventually, it should be possible to replace more than half of the human body with artificial organs."[45]

Our inventory of spare parts is growing at a dramatic rate, and it is not hard to see why. Natural parts are in short supply, but we could manufacture as many artificial versions as we wanted. We could make them impervious to rejection, infection, or degenerative diseases, and perhaps we could even build in features that give the wearer new powers. But there also will be drawbacks, especially for artificial versions of complex organs. We will find they are usually more expensive and less versatile than the natural part. For example, they will lack the original part's variety of functions, its adaptability to stress and growth, its capacity for self-repair, and its ability to use the body's own energy system.

Let us scan the catalog, beginning with devices that help restore functions of the limbs. Artificial thigh bones made of titanium and other alloys are being developed. We can use dacron to repair or replace tendons. Arm and hand prostheses are becoming so sophisticated that with some models the wearer can control the device subconsciously, using electrical signals from nerve endings in his stump. Some models even provide a sense of touch and pressure. Artificial lower legs also are in use. The ultimate goal is for man and his prostheses to be integrated into a single unit; he will control his artificial limbs the same way he controls his natural ones—mentally.

Replacing worn-out joints is now a reality. The greatest success story is replacing the hip with a ball-and-socket joint of stainless steel (or some other alloy) and plastic. Once cemented in place, the artificial hip performs many of the motions of the natural hip. These devices have helped tens of thousands of people, many of them suffering from arthritis, to resume their normal day-to-day activities. Replacement joints are also coming of age for the wrist, fingers, elbow, shoulder, ankle, and knee.

The circulatory system is also getting some technical assistance. Many thousands of synthetic heart valves and electronic pacemakers help hearts carry out their normal functions. Over a hundred thousand plastic artery segments have been installed in people with damaged blood vessels. Blood substitutes are available for temporary use. Tens of thousands of people have received silicone rubber tubes that pass from the brain cavity, through the skull, and into the jugular vein. These tubes drain excess cerebrospinal fluid into the bloodstream, relieving the pressure of hydrocephalus ("water on the brain").

Increasing numbers of people are regaining the use of their senses, thanks to modern technology. Artificial teeth have been with us for many years, and now scientists are trying to develop a way to implant them directly in the jawbone. Several types of artificial larynxes are in use. Artificial eardrums may replace perforated eardrums. Cosmetic prostheses can replace deformed ears, noses, and other body parts. And it may even prove possible to restore hearing to the totally deaf by converting sounds into electrical impulses, and then sending those impulses directly to the deaf person's brain for translation. A similar approach might work for restoring vision. Indeed, there have been several promising experiments with blind people where light signals were converted into electrical impulses that were transmitted by a matrix of electrodes to a sensitive area on the person's skin, or even sent directly to his brain. Both versions require him to recognize a dot pattern that is "like a picture on the scoreboard of the Houston Astrodome."[46]

People with failing organs also should consult their friendly mechanic. If their problem is kidney failure, their toxic wastes can be removed by circulating their blood through a dialysis machine, which consists of semipermeable, cellophanelike membranes bathed in a nutrient fluid. There are now portable, eight-pound units that can be strapped to the chest, and implantable units are on the way. One of the most promising ideas is CAPD (continuous ambulatory peritoneal dialysis), which eliminates the machine entirely. Here the patient pours the dialysis fluid through a permanently implanted tube into his abdomen; blood vessels in the peritoneal cavity give up their wastes to the dialysis fluid, and the fluid is poured out through the tube after a few hours. Tests so far indicate CAPD may be a cheaper and more effective way for many of the thirty-six thousand Americans who depend on artificial kidney machines to stay alive.[47]

People with liver failure may have toxic materials in their blood that dialysis does not remove. One solution may come in the humble form of charcoal, whose ability to adsorb impurities has found many uses, including cigarette filters and gas masks. Indeed, physicians have detoxified patients with severe liver disease by circulating their blood through a chamber containing granules of plastic-coated charcoal.[48] Another possibility is to circulate their blood through a membrane that has liver enzymes from animals on the other side; the liver enzymes then metabolize the toxins.[49] These experimental methods could buy time for the patients, giving their own livers a chance to recover. But the liver is the Grand Central Station of the body in terms of chemical activities, so there is no realistic chance of inventing an artificial liver that could permanently replace the real thing.

A better candidate for replacement is the pancreas. One of its main functions is to keep the blood sugar (glucose) levels within a narrow range. When those levels rise the pancreas normally secretes insulin, which reduces the amount of glucose in the blood. Now scientists are developing a device that has a sensor to monitor glucose concentrations and a tiny computer to process the information and activate a pump that injects either insulin or glucose, as the situation requires.[50] Although a few experiments with diabetic patients have been encouraging, implantable units are not yet in sight. Another idea is to implant pancreatic cells that are embedded in a plastic coil, which helps protect the cells against rejection.

The prospects for an implantable artificial lung are dim, for few materials can match the natural lung in letting gases pass in and out of the body. But temporary assistance is quite a different matter. The respirator is a machine that in essence breathes air in and out of the patient's lungs. Another device, the membrane oxygenator, bypasses the lungs entirely; a complete unit can circulate blood out of the body, add oxygen, remove carbon dioxide, and then return the blood to the patient. Though membrane oxygenators are at best used for only a few days at a time, respirators can keep people alive almost indefinitely.

We are developing an impressive array of devices to assist the heart and to take over its functions temporarily, but the most dramatic prospect is an implantable, artificial heart. It would be an electric- or nuclear-powered pump made of plastic, silicone rubber, or natural rubber. Already, a bull calf has lived 210 days with an implanted, polyurethane heart; his heart was still working when he died, but he had outgrown its pumping capacity.[51]

Only one human being has had an artificial heart placed in his body. In 1969 Denton Cooley of the Baylor University College of Medicine performed the implantation in an effort to buy his patient time for a heart transplant. The patient survived sixty-four hours on the artificial heart, then received a heart transplant but died thirty-one hours later. By 1977, however, at least twenty-two patients had been connected to mechanical pumps as last resorts to save them after open-heart surgery. The twenty-second time was the charm. A thirty-nine-year-old woman was supported for eight days by an external artificial heart until her heart was able to take over. After four months of hospitalization she went home.[52]

Overall, the future is bright for transplants and artificial organs. But there are limits to the benefits we will receive. Alex Comfort has explained: "A law of diminishing returns operates in purely palliative life preservation, and at great ages cure of one disorder merely exposes, and sometimes aggravates, another—rather as replacement of a faulty component in an old radio may restore voltages to their correct original levels and blow out several other components which can no longer stand them."[53]

Unless only a few organs limit our life span, and unless we can effectively replace all of them, transplants and artificial organs will not increase our maximum life span. One of those limiting factors is the brain, and until we learn how to revitalize that organ for a few extra decades or centuries of life, it is pointless to try living beyond our current span. In practice, then, we will hardly come to use replacement parts as a form of preventive maintenance, like changing the oil filter on a car every three thousand miles.

Nevertheless, transplants and artificial organs will increase our life expectancies, and even a modest increase will have a major impact on our society. These technologies also raise important ethical and legal issues. Indeed, we will find that the ultimate limits of replacing body parts will not come from our technology but from our value judgments of what is most desirable. Now let us consider some of those issues.

SOME CONCERNS

Donating Organs

As our technology advances, and the average age of our population rises, we will find increasing numbers of people who would benefit from replacement parts. But where will those parts come from? For people needing a complex organ, the best solution will usually be a transplant. Animal organs will generally be used only as a last resort, so the recipients will have to depend on other people to furnish those organs. But that raises several ethical and legal problems.

First, let us consider the issues with live donors. Many people believe a person's body is a gift from God, and some interpret this to mean that no one may choose to be mutilated (as in donating an organ) except when it is for the welfare of his own body. Indeed, France has prohibited surgery that is not for the benefit of the patient, and this has had the effect of prohibiting transplants from live donors.[54] On the other hand, surely the well-being of the total person, including his psychological and spiritual health, is more important than his physical health alone. Otherwise, it would be immoral for people to have vasectomies or cosmetic surgery, even though they considered it to be in their own best interests. And it would be immoral for anyone to endanger himself to help others—for example, trying to rescue a fallen mountain climber, protecting family members from an attack—or donating an organ.

People should have the right to decide for themselves whether the benefits of donating an organ outweigh the risks. They must weigh the risks to their own health, such as the risk of dying from an operation to remove a kidney, or from being left with a single kidney. They also must consider the psychological effects. After surgery, some donors feel depressed; they resent the recipient and the people who encouraged them to donate. But those feelings usually disappear, and in the end most donors feel much better about themselves, even if the transplant fails.

The risks are real, nevertheless, so it is important that each donor truly give his voluntary and informed consent. This high, but necessary, standard would generally exclude minors, mental incompetents, and people such as prisoners who are under duress. It is also hard to ensure that a donor, regardless of age, truly gives his consent voluntarily when he is related to the recipient. For example, he may "volunteer" in order to avoid feelings of guilt or to demonstrate his loyalty and courage. Because of these problems, hospitals often use procedures such as the following: A psychological evaluation is given to assess, however imperfectly, whether the donor is mentally stable and whether he is being coerced; the evaluation is done before the recipient enters the hospital, and by people who are not involved with the care of the recipient; the donor is fully informed about the risks to himself, the chances for a successful transplant, and other medical options; he is given ample time to consider his decision; and he is told that if he decides not to donate, the physician will provide a technical excuse if necessary.[55]

People do not always decide to donate. In 1978 a thirty-nine-year-old man dying of a rare disease (aplastic anemia) needed a bone-marrow transplant. The only person who was suitably matched was his cousin, but he declined to donate. The patient went to court, asking that his cousin "aid a dying man [through] a medically safe, experimentally proven, minor procedure which will at most result in minor and temporary discomfort." The judge denied the request on the grounds that it "would defeat the sanctity of the individual." The man died two weeks later.[56]

The complications of using live donors are great, and the types of organs they can supply are limited, so we will have to look elsewhere for most of our replacement parts. Which brings us to the only other source of human organs—cadavers.

Our growing dependence on cadavers raises another set of problems. We might imagine, for example, new opportunities for grave robbers and body snatchers. But organs deteriorate so rapidly that the culprits would find a rather small market for their wares. Indeed, it is safe to assume that a "hardhearted person" will remain only a figure of speech.

Some people believe it would be disrespectful to the deceased if the corpse were disturbed by removing a few tissues. And some believe this would be sacrilege because the corpse is waiting to be resurrected and reunited with its spirit. While this clearly is a matter of personal belief, it is by no means clear why only an intact body could or would be resurrected. According to one estimate, however, "the number of people who believe they will need both their kidneys in the hereafter is comparatively small."[57]

The Uniform Anatomical Gift Act, a piece of model legislation that has been substantially adopted by all fifty states, allows any mentally competent person of eighteen years or older to donate any or all of his cadaver to any hospital, accredited medical or dental school, university, organ storage facility, physician, or individual for use in scientific research, education, therapy, or transplantation. He can make the donation in his will, or by a signed, written statement in the presence of two witnesses. No one can overrule the donation. The law also allows the next of kin to authorize donations unless there is evidence that the decedent would have objected.

Despite this legislation, we still do not have enough organs for transplantation. Indeed, thousands of people in the United States alone are currently waiting for a suitable kidney. The reasons for the organ shortage are a bit elusive. According to a Gallup poll, most people (70 percent) are willing to donate organs, and their willingness does not correlate significantly with their religious affiliation.[58] Perhaps some people fear they would receive less complete medical care when they approached death if they were known to be potential donors. Yet the law specifies that the physician who certifies death may not be involved with the transplant. So that leaves us with a rather simple explanation for much of the shortage—apathy and ignorance.

How could we increase the supply? One answer is to inform people about the seriousness of the problem and the simple procedures for making donations. In blunt terms, the message "is simply that people dying of kidney diseases or suffering from blindness deserve to be given priority over corpses."[59] Making it easier to donate also might help. Indeed, forty states now allow residents to "will" their kidneys by a notation on their driver's license. But this has had little impact on the organ shortage, for the document is not legally binding and the next of kin must still consent to the removal. So while the educational approach will help, it is a poor bet to solve the problem.

There are several legislative approaches. One option is to let people sell their body parts. Indeed, in 1977 a twenty-one-year-old Brazilian girl with two extra kidneys said she hoped to sell them for about thirty thousand dollars each.[60] People in the United States can already sell their blood and sperm, so this approach would simply extend that practice to include other body parts and allow later delivery of the promised goods. But the drawbacks could be substantial: a black market, with organs being sold to the highest bidder; poor quality control; and assorted ways to exploit the poor. And since corpses would have commercial value (at least for an hour or so), we might even witness the bizarre eighteenth-century spectacle of corpses being arrested for not paying off their debts.

Another option is to require corpses to surrender parts that someone needs for transplantation. Here we could argue that corpses have no rights and the state has an overriding duty to protect the health of its citizens. Compulsory "donation" certainly would increase the supply of organs, but the restrictions on individual freedom—especially the right to hold and observe religious beliefs—would probably be unconstitutional in the United States.

A third approach, one that is used in Sweden, Israel, Italy, and France, is to presume people are willing to donate body parts at the time of death unless they, or their next of kin, have filed an objection. This transfers the burden of action to the minority (according to public opinion polls[61]), and saves people who are awaiting a transplant from being the victims of someone else's apathy. This arrangement also preserves each person's rights, provided there is an efficient method for filing objections and ensuring they are honored. Indeed, the benefits are so compelling that the United States should consider adopting this type of system.

Choosing Recipients

Supplying replacement parts is only part of the problem; we also have to decide who will receive them. With natural organs in such short supply, the task is fairly straightforward with transplants. First, patients become eligible if they meet certain medical criteria such as age (generally, sixty years or less) and health. They may also receive a psychological screening to assess whether they can deal with possible problems—feelings of guilt toward a live donor, an impaired body image, an identity crisis. According to Denton Cooley: "Not everyone wants to live with someone else's heart inside him. It's an eerie thing."[62] Then, from the pool of eligible candidates, the actual recipients are chosen simply on the basis of whoever happens to be the best match for an organ that is available. The selection is done by a computerized antigen-matching system.

With artificial organs, however, the decisions are not as clear-cut. For one thing, the psychological adjustments may be more difficult. Not everyone can accept a regimen of twenty to thirty hours every week connected to a dialysis machine, or the reality that his life and activities depend on a machine. And people with external devices, such as an artificial arm or larynx, must cope with sometimes being considered less than a full person. Improved design will ease some of these problems, but this can only be part of the answer. For until we are able to interact comfortably with others who bear some mark of abnormality, their psychological problems will be compounded by ours.

Even after medical and psychological criteria are applied, there are still situations where there are not enough artificial organs for the people who qualify. This problem has arisen with people who need kidney transplants; they are maintained on "artificial kidney" machines until a natural kidney becomes available. But that can take a long time, and there are a limited number of machines at a dialysis center, so sometimes a decision must be made on who will get to use a machine.

How do we make such life-and-death decisions? One method has been to assess "social worth." Some hospitals have used a committee of physicians and, perhaps, other community members to select the winning candidates. The committee members evaluate such factors as occupation, marital status, number and age of dependents, life expectancy, social standing, and contributions to society. Contributions to the hospital presumably do not count.

This approach has serious drawbacks. One is the suspicion (even if it is unjustified) that political factors influence the decisionmaking. Another is the basic belief that no one can validly judge the worth of another human being. These concerns have led to other, nonjudgmental methods. The most common is first come, first served; random selection is another. While it is not at all clear these methods are better, it is just as well they are being tried. For the "social worth" approach has produced an alarming decline in the number of philosophers.

Since we do not have the resources to supply everyone with every replacement part, we must also weigh costs against benefits. For complex organs, transplants will usually be cheaper and more effective than mechanical devices; when only one part of an organ is damaged, or a body part is mostly mechanical, artificial substitutes will often be better. But the problem of cost is especially great with artificial parts. The Food and Drug Administration, which regulates medical devices, upholds high standards of safety, effectiveness, and quality control. This increases the cost of developing and testing new products, so companies often find that if the potential market is small, they have little chance of making a profit unless they receive a subsidy. Indeed, there is at least one instance where a private company declined to make a needed medical device because the cost of doing the necessary tests far outweighed the potential sales.[63]

As our artificial parts become more intricate, we will reach a point of diminishing returns, where we decide certain features just are not worth the extra cost. We may have to set both minimum and maximum requirements for medical devices, and each "extra" will have to be justified on a cost-benefit basis.

We also have to weigh the costs and benefits to society as a whole. Everyone may be morally entitled to be a six-million-dollar person, but we simply do not have the resources to do it. One physician even predicted: "The pattern of advance of medical technology suggests that before too long we may have to decree that the various pumps, potions and prostheses which can keep a man alive beyond his natural span should be withdrawn when he reaches some statutory age."[64]

While that is a bit harsh, we cannot ignore the cost of hastening our march toward an older and larger population. Already, 10 percent of the people in the United States are sixty-five or older, and this group is increasing at a rate three times faster than the population as a whole. And the trend is world-wide. According to one speculation, the people in the world who are sixty-five or older now constitute nearly one-fourth of all the people who have ever reached that age.[65]

The social impact of this trend is enormous. One challenge is to help old people retain a major role in society. They are a valuable resource of wisdom and experience, but they often feel left behind as their physical abilities wane and they are confronted with a dizzying rate of change in society. As Peter Medawar, a British immunologist and Nobel laureate remarked: "Today the world changes so quickly that in growing up we take leave not just of youth but of the world we were young in."[66]

So far, we have not helped the aged very much. In fact, we have tended to increase their problems by segregating them—both physically and psychologically. Many decide to live in retirement homes; nursing homes are another answer—especially for their relatives. Old people living in separate facilities have ready access to medical care, but they have a more important need: human care. Indeed, our emphasis needs to be on the quality, rather than the quantity, of the later years of life.

We must also face the prospect of a shrinking proportion of young people having to support an increasing proportion of old people. One critical area will be medical services. People sixty-five and older already account for nearly one-third of the health costs in the United States, and this proportion will increase. Furthermore, as more people come to believe health care is a basic human right that should not depend on a person's ability to pay, we will commit increasing amounts of our public resources to health care. That money will have to come from somewhere: perhaps higher taxes, or less money for such areas as law enforcement, education, housing, national defense, and social services. But whatever we do, we will still reach a limit on how much we can spend on medical care, and we will have to choose our priorities within that budget.

Here we must acknowledge a harsh reality: As our physical and mental abilities dim, we become increasingly poor investments for those limited resources. George Pickering, professor of medicine at Oxford University, remarked: "The present goal of medicine seems to be indefinite life, perhaps in the end with somebody else's heart or liver, somebody else's arteries but not with'somebody else's brain. Should transplants succeed, those with senile brains will form an ever-increasing fraction of the inhabitants of the earth. I find this a terrifying prospect."[67]

That does not have to happen; we need not squander our resources on patchwork or bionic Mister Magoos. Fortunately, we have a simpler and more satisfying answer, both for the individual and for society as a whole: We can come to accept, with grace and dignity, death.

Life and Death

One physician has written: "Doctors can keep almost anyone alive artificially nowadays. That's the measure of medical progress in the second half of the 20th century."[68] Indeed, our hospitals now have an impressive array of hardware to keep us going. As temporary measures, respirators, heart-assist devices, and other machines have helped save many lives. But they also force us to make agonizing decisions about when to use them.

Machines can keep a patient hovering between life and death and they can even mask his death. For as long as they circulate, oxygenate, and cleanse his blood, his organs will continue many of their normal functions. That makes it hard to determine when a person is dead. Now we are turning away from the idea that the heart is the crucial organ; instead, we are coming to believe that a person dies when his brain dies, even if machines can sustain the rest of his body. In 1968 a committee of thirteen members of the Harvard Medical School staff proposed specific medical criteria for defining "brain death."[69] That medical definition has now been accepted by most physicians, and it was upheld in 1977 by the Massachusetts Supreme Court.[70] And eighteen states have passed laws defining death as the loss of brain function. [71]

But what about patients who do not meet that definition, people who are slowly dying and have only a small residue of brain-wave activity? When is it appropriate to turn their machines on, or off? People throughout the world pondered this question as they read of the tragic events surrounding Karen Quinlan and her family.

On the night of April 15, 1975, for reasons that were never fully determined, twenty-one-year-old Karen Ann Quinlan went into a coma. She was rushed to Newton (New Jersey) Memorial Hospital and placed on a respirator. Nine days later she was transferred to another hospital. As weeks and months passed by, she remained on the respirator unconscious. She had short periods of spontaneous breathing; her eyes and mouth occasionally opened; her eyes sometimes followed the source of sounds; and she had some brain activity. She did not meet the medical criteria for death, but no doctors held any hope that she would recover. Her body gradually curled into a rigid, twisted state, and her weight dropped.

After several months Karen's parents asked that the respirator be removed. The attending physicians declined on the basis that such action would be illegal and contrary to the medical code. Then her parents asked the New Jersey State Court to designate the father as guardian so he could authorize that the respirator be removed. Their request was denied. But in March 1976 the New Jersey Supreme Court unanimously granted Joseph Quinlan the request under the condition that competent medical authorities agreed Karen had no reasonable chance of recovery. It was done; in May Karen was taken off the respirator. To almost everyone's surprise, she was able to breathe on her own. As of January 1979 she was still alive and unconscious.

Her case stirred people to think carefully about our life-sustaining machines and the problems in setting limits on their use. The answers are not easy, and there are many factors to consider. Friends and relatives feel great distress while a loved one teeters indefinitely in the misty no-man's-land between life and death. Sometimes they are also drained financially. While Karen Quinlan was on the respirator, for example, her total expenses exceeded one hundred thousand dollars. But in her case, and many others, the expenses are borne primarily by the public. Now we could argue that, as a matter of public policy, the most productive use of our medical resources is to reserve them for patients who have a reasonable chance of recovery. Yet few people would go so far as to insist, in the name of the public good, that individuals may not cling to a thread of life, however slender.

One answer is to let the family decide. For example, the Medical Society of New Jersey has proposed that each hospital establish a committee capable of assessing whether borderline patients are likely to function again at a reasonable mental level. The committee would advise the family and, if the prognosis were negative, the family could choose whether to have the life-support system removed.[72]

A physician may argue, however, that his first and foremost duty is to preserve his patient's life. But preserving life is not necessarily the same thing as prolonging dying. Furthermore, as one physician remarked: "The obligation of the physician is to do the best he can for the patient. And sometimes the best thing he can do is let the patient die."[73] When a patient's mind is fading and his capacity for self-reflection and human relationships is lost irretrievably, according to competent medical judgment, insisting that his physical life go on may be more cruel than loving. There is a profound difference between living and existing.

Others may protest that life is a gift from God and, therefore, man has no moral right to terminate it. It does not necessarily follow, however, that we should use machines to prolong life, even when there is no reasonable chance of recovery. Over two decades ago, Pius XII said that "extraordinary" means were not required to prolong life. And the Church of England has stated that life-support machines may be withdrawn from a patient if doctors determine that his organs will never again function on their own. Nevertheless, some people believe that pulling the plug is immoral because it frustrates God's will for us to live. That argument, however, rings a bit hollow in the face of the plea from Karen Quinlan's parents: "Take her from the machine and let her pass into the hands of the Lord. . . . If he wants her to live in a natural state, he'll create a miracle and she'll live. If he wants her to die, she will be off all the artificial means and she'll die whenever he calls her."[74]

It's an awesome responsibility to decide whether to withdraw life-support systems from someone else; a better arrangement is to have the person decide for himself. But does he have the legal right to do so? A New York court ruled in 1914 that "every human being of adult years and sound mind has a right to determine what shall be done with his own body."[75] Yet in another case, where a man refused an operation and consequently was waiting to die, a justice stated: "Three months ago the medical men knew this would happen, but according to our present legislation, or lack of legislation, they were helpless. The poor immigrant, through ignorance or foolhardiness, or both, forbade them putting forth the hand which would have saved his life. Should not society protect such a man from his own foolishness?"[76]

The answer must be a qualified no. First of all, people who decide to forego medical treatment are not necessarily "foolish." Nor are they necessarily incompetent; indeed, such decisions may often be saner than those to endure awhile longer. Furthermore, as long as a mentally competent person is fully informed (or has refused the information) and is acting voluntarily and without malice, he should be free to make decisions that someone else may consider foolish. After all, we protect the rights of consenting adults, in private, to smoke, read philosophy, and even watch television.

People on the verge of death, however, are not always competent to make binding decisions. So some people have signed "living wills," specifying in advance that extraordinary means should not be used to keep them alive. Those documents may not be legally binding except in states that have adopted a living-will statute. California was the first to do so (in 1976), and other states have followed suit. As Gordon Rattray Taylor, a science writer, remarked: "The attempt to die, despite the efforts of those who would keep one alive, might form the most macabre of dramas, and the right to die may one day need to be defended as the most fundamental of human liberties."[77]

It is about time. Living-will statutes will not set off a stampede of customers, or cause instant bankruptcy for life insurance companies. But they will encourage many of us to think realistically about death, and to develop some attitudes and answers we literally can live with. Philip Abelson, editor of *Science,* has written: "Death of a loved one was bad enough when it was in the hands of God; now it is often a much more distressing experience."[78] Yet it does not have to be that way. Indeed, the very machines that threaten to keep us alive indefinitely also challenge us to do something about it—to free ourselves from an inordinate fear of death.

We need not cringe at the prospect of death; it is a natural and necessary part of life. Although in the twentieth century we have largely removed death from the home to the hospital, we are coming again to realize that the most important need of a dying person is to be with his family and friends. One answer is the "hospice," a concept of care that enables the patient to live as normal a life as possible. Whether he is in his home, a hospital, or some other facility, the dying person receives whatever pain-killing and anxiety-relieving drugs he needs, and he is with people who care for his other needs—relatives, friends, and specially trained health personnel and social workers. He is free to talk about death, and to share his experience with those around him.

Indeed, we are learning a great deal from the terminally ill. The major pioneer has been Elisabeth Kübler-Ross, a Swiss-born psychiatrist who has encouraged thousands of dying patients to talk about their experiences. She has also interviewed people who have returned from the brink of death, and their experiences are comforting. She has reported: "They have a fabulous feeling of peace and wholeness. . . . People who are blind can see, paraplegics have legs that they can move. They have no fear, no anxiety. . . . In fact it is such a beautiful experience that many resent being brought back to their physical body."[79]

If we truly respect human life, we cannot insist that everyone live as long as modern technology can keep his body sort of functioning. To the contrary. We should help people feel the rich fabric of life, but we also should respect their decisions to decline further treatment when life has become only a burdensome existence. We must learn how to make death a decent and humane experience, not something bitterly contested to the last gasp. Paul Ramsey has termed death "the ultimate indignity."[80] Living will proponents often speak of "death with dignity." But the point of it all is to enhance human dignity, not particularly in death, but in life.

I don't believe I am being melodramatic in suggesting that what our research may discover may carry with it even more serious implications than the awful, in both senses of the word, achievements of atomic physics. Let us not find ourselves in the position of being caught foolishly surprised, naively perplexed and touchingly full of publicly displayed guilt at what we have wrought.

David Krech

Special hormones or other chemical agents will be used to reinforce the vigour of a man's mind, to strengthen his character, to dispose him to virtuousness. Quite soon, perhaps, people will buy genius or sanctity at the chemist's, just as now women buy the straightness of their nose or the depth of their gaze at the beauty parlor.

Jean Rostand

8

New Minds

As I write this sentence, it is fascinating to reflect on what must be happening inside my brain. The idea I want to communicate must be put into some type of language, so from the myriad words of the English language stored in my brain in some mysterious form, I must select a few and arrange them according to the conventional rules of sentence construction. My brain has to decide, from the almost limitless possibilities, the best combination to express my idea clearly, accurately, and with the right emotional impact. As my brain commands my arms and fingers to convert those words into written form, my eyes send signals back to my brain so I can detect any mechanical errors. I also use those signals to analyze whether the sentence is all right, or whether I should change it. This often takes longer than everything that preceded it.

It is amazing we can do all that and do it so quickly. Our brains can receive, process, and retrieve information so rapidly that we can do things virtually the same time we think about them. In fact, some of us are so fast that on occasion our words and other actions even precede our thoughts.

More than anything else, the brain makes man man. It gives us a unique and dominant status in nature, for we have a powerful ability not only to adapt to a changing world but also to mold that world to fit our needs. In addition, our brains underlie our identity as persons, making us conscious, thinking, moral, and emotional beings.

The brain is central to our humanity, but its workings continue to mystify us. Nevertheless, scientists are acquiring an imposing array of tools to probe and manipulate the brain, so we may well be entering an era of dramatic new understanding. In this chapter we will explore some of those tools and their potential impact on us.

THE BRAIN

An adult human brain is about one and a half quarts and three pounds of pinkish-gray material, roughly the consistency of soft cheese. It has ten billion nerve cells (neurons), which transmit nerve messages, plus about ten times as many glial cells, which nourish and support the neurons. A newborn's brain is about one-fourth of its adult weight, and it has virtually all the neurons it will ever have (though some brain cells—probably glial cells—may be added in the first year). By age six the brain is 90 percent of its adult size; later in life it begins to shrink as neurons are permanently lost.

We can think of our brains as having three layers. The bottom, innermost region is the brain stem, which connects the brain with the spinal cord, thus establishing communication with other parts of the body. The brain stem functions like a discriminating switchboard operator, monitoring signals between the brain and the rest of the body, rejecting those that are irrelevant or trivial, and controlling the attention level of the brain to messages that do come through. So the brain stem can arouse the rest of the brain for action, or let it drift off to Wonderland.

The middle layer has several structures—including the hypothalamus, thalamus, pituitary, hippocampus, and amygdala—that are interconnected by an extensive nerve network. This region controls most of our hormone activity and is responsible for our moods, emotions, and basic drives. In addition, it helps monitor and adjust our heartbeat, blood pressure, temperature, and other involuntary functions.

The outermost layer is the cerebral cortex (cerebrum), the wrinkled surface of the brain. Parceled into two distinct hemispheres that communicate with each other through millions of neurons, the cortex takes up about 80 percent of the brain's volume. In terms of evolution, the cortex is the part of the human brain that has developed most in size and most dintinguishes us from other animals. It is largely responsible for our senses, voluntary muscle control, abstract thought, memory, language, creativity, and consciousness.

Beneath and behind the cortex lies the cerebellum, a fist-size structure that helps monitor and coordinate our voluntary muscular movements.

Scientists are trying to unravel the detailed structures and activities of each part of the human brain and the mechanics of how those activities are carried out. But there are two reasons why progress has been slow. First, and foremost, the workings of the human brain are incredibly complex. Second, there are few experiments scientists can safely do on normal, functioning human brains, so most new information must come from experiments with animals, or from attempts to treat people with brain disorders. Nevertheless, scientists have developed some remarkable tools to alter the brain and probe its secrets. Now let us examine several techniques.

ALTERING THE BRAIN

Transplantation
Although we will learn how to replace most of our worn-out or diseased body parts, we would not use that technology to live beyond one hundred years or so unless we could also rejuvenate or replace our brains. One idea comes from James Bonner, who speculated: "Someday it may be possible to replace, through synthesis of brain neurons, the 100,000 neurons in the brain that die every day. The new cells will have to be trained, because they will have no memory."[1] So this dim and distant prospect would pose a new set of problems.

An even more radical "solution" would be to replace an entire brain by transplantation. Robert White, professor of neurosurgery at Case Western Reserve School of Medicine, has developed techniques for connecting a second brain to blood vessels in the neck of a host animal. By 1968, he had transplanted dog brains in this way and found they had metabolic and brain wave activity for as long as three days.[2] Then he transplanted monkey heads onto headless monkey bodies; their brains functioned up to seven days, with the monkey eyes following stimuli and the mouths chewing food.[3] This improved on work done much earlier by a Russian surgeon, V. P. Demikhov, who had reported a survival time of one to twenty-nine days for the upper body (head, neck, and upper limbs) transplants of puppies onto the neck region of larger dogs. Some of those transplanted heads followed the movement of people in a room, licked milk from a bowl, and even bit the head of the host dog.[4] That is gratitude for you.

Since the brain is fairly inaccessible to the immune system, a transplanted brain might elude a common problem of transplants—rejection, even if there is no close tissue match. And although scientists have had little success in storing complex human organs, perhaps they will eventually succeed, even with the brain. Indeed, in one experiment a cat's brain that had been surgically removed, treated with preservative, frozen, stored for six months, thawed, and supplied with fresh blood then resumed its brain-wave activity.[5]

These techniques are valuable research tools, and they have revealed important insights into the workings of the brain, but they have hardly brought us to the verge of human brain transplants. Metabolic and brain-wave activity in animal experiments hardly assure us that a thawed or transplanted human brain would work satisfactorily. Furthermore, nerve cells have a poor ability to regenerate, so severed nerves from a transplanted brain probably could not establish normal communication with its new body. Nor would a transplanted head be able to communicate its thoughts by speaking. Indeed, White said in 1977 that within one year he would technically be able to transplant a human head, but the paralyzed body would serve simply as a life-support system for the brain.[6]

I have not heard of anyone volunteering for that experiment. Some people may think we should use this method to save brilliant minds that were encased in a degenerating body. Perhaps "heads of state" would take on a whole new meaning. Still, few brains would choose such an existence. And even if they did, we might question whether they were the type we wanted to preserve. As far as the benefits to society are concerned, I suspect we will somehow manage to get along despite losing a brilliant brain now and then.

Perhaps it is just as well human brain transplants are not practical, for the problems would outweigh the benefits. Some of us have enough trouble with self-confidence as things are now. Think of the complications if a person literally lost his head. He could have most parts of his body replaced and still feel he was basically the same person. But not his brain. Which is why there actually is no such thing as head transplants; they are body transplants.

Psychosurgery

Wilder Penfield, an eminent Canadian neurosurgeon, was a pioneer in using brain surgery for therapy and research. In the 1930s he began operating on people with head injuries or severe epilepsy, and he also used many of those occasions to learn more about their brains. The results were startling. When he stimulated certain regions of the cortex while his patients were awake, a few of them suddenly and vividly recalled specific experiences. And stimulating an area at the back of their brains produced visual sensations his patients could not integrate into a whole picture. They reported: "flickering lights, dancing lights, colors, bright lights, star wheels, blue, green and red colored discs, fawn and blue lights, colored balls whirling, radiating gray spots becoming pink and blue, a long white mark, *et cetera.*"[7]

Other researchers were probing animal brains and discovering that surgery could produce dramatic changes in behavior. As a result, some thought surgery might be an effective way to treat mental disorders in people (an approach now called psychosurgery). Egas Moniz, a Portuguese neuropsychiatrist, performed the first such operation—a lobotomy—in 1935. From then until the 1950s, lobotomies became a common way to treat mental disturbances, especially schizophrenia. There are various versions of this operation, but the key feature is to cut brain tissue so as to disconnect part of the cortex, usually the frontal lobe, from the area with the hypothalamus. In the wave of public enthusiasm for this method, tens of thousands of people received lobotomies—with mixed results. While some improved, others gradually deteriorated. Some lost their imagination and motivation and no longer cared about their own behavior; indeed, they seemed to lose their souls.

Lobotomies are no longer popular, for they have not proved effective enough to outweigh the risks, especially in comparison with drug therapy. But two other methods of brain surgery have proved useful, both for therapy and research. One is the hemispherectomy, where the surgeon removes one hemisphere of the brain. This is usually done because there is a tumor growing in that hemisphere. From studying those patients after the operation, scientists have learned about the roles of the two hemispheres. They have discovered that in most cases the left hemisphere controls verbal expression while the right hemisphere is mute. Occasionally the relationship is reversed, or both sides have verbal abilities. Age is also a factor, for adults who lose their left hemisphere typically lose almost all their ability to speak and write; children, however, generally regain their verbal ability, perhaps because of some compensating capacity in the right hemisphere.

Another tool is the split-brain operation, where the surgeon severs the major nerve cable between the hemispheres. This helps epileptics by reducing the chances that abnormal electrical patterns will spread across their brains and set off seizures. Although the operation seems radical, split-brain patients function normally. But they cannot integrate the activities of their hemispheres, and this produces unusual effects in special exercises. Since each hemisphere controls and interacts with the opposite side of the body, there are exercises where information can be given selectively to one hemisphere. Roger Sperry of the California Institute of Technology, the leading authority on split-brain experimentation, has described how it works:

For example: the subject may be blindfolded and some familiar object like a pencil, a cigarette, a comb or a coin placed in the left hand. Under these conditions the mute [right] hemisphere connected to the left hand feeling the object perceives and appears to know quite well what the object is. Though it cannot express this knowledge in speech or in writing, it can manipulate the object correctly, it can demonstrate how the object is supposed to be used, and it can remember the object and go and retrieve it with the same hand from among an array of other objects either by touch or sight. While all this is going on, the other hemisphere meanwhile has no conception of what the object is and, if asked, says so.[8]

In other experiments, where the hemispheres receive different information, a split-brain person may simultaneously attempt two different and even contradictory tasks.

From studying people with hemispherectomies or split-brain operations, scientists have discovered that the two hemispheres function quite differently. The current picture is this: The dominant hemisphere, usually the left, has the major role in verbal, analytical, sequential, and computational tasks, while the other hemisphere deals mostly with spatial, musical, and pictorial tasks and helps organize individual impressions into general patterns. Art is primarily a right hemisphere activity, for example, while bookkeeping is associated more with the left. Yet we should not overstate those differences, for the hemispheres normally communicate with each other and integrate their individual activities; furthermore, each has some ability to take over functions of the other, especially in the early years of life.[9]

Whereas brain surgery has provided valuable insights into the workings of the brain, its therapeutic value has been modest. Split-brain operations and hemispherectomies involve only a few people, and the results of lobotomies are less than inspiring. Until we understand the brain in much greater detail, and until surgeons have more precise techniques that destroy as little tissue as necessary, psychosurgery will remain a risky and crude way to treat mental disorders. That knowledge is still far beyond our grasp, but we already have a more precise tool for manipulating specific bits of brain tissue—electrical stimulation by implanted electrodes.

Electrical Stimulation of the Brain (ESB)

We have come a long way since Wilder Penfield began stimulating brains during his operations. Now scientists can send electric impulses through metal electrodes, one-millionth of an inch in diameter, that are implanted in individual nerve cells in the brain. And they can send the impulses by remote control. They can even program a computer to control the delivery of a preset pattern of impulses, or to alter the pattern depending on the electrical patterns the brain itself is producing. The possibilities are awesome.

One use of ESB is to map the brain. Scientists have discovered, for example, that stimulating an area about one inch wide in the cortex of each hemisphere will cause muscular movements. With ESB, they can map this motor strip to correlate specific regions with specific actions. An electric impulse to a particular region might cause a person to clench his fist; he cannot mentally overrule that action, and he might not even be aware of it.

Jose Delgado of Yale University, one of the leading (and most controversial) practitioners of ESB, has demonstrated some remarkable control over motor movements. When he stimulated a particular region of one monkey's brain, the monkey would stop whatever it was doing and go through the following sequence: make a face, turn its head to the right, stand up, circle to the right, walk on two feet, climb the cage wall, come down, grunt, glare at the nearest monkey, become peaceful, and resume normal activity. In response to ESB, that monkey went through the entire sequence of actions once a minute for two weeks, a total of twenty thousand times.[10] For his most famous exhibition, Delgado used a bull with electrodes implanted in a region of the brain that blocks motor activity. Armed with a radio transmitter and a red cape, he stood in a Spanish bullring as the bull was released. When the charging bull was a short distance from him, Delgado pressed the button on the transmitter and brought the bull to a screeching halt.

ESB has also revealed regions in the brain that harbor our emotions and basic drives, such as hunger, thirst, and sex. For example, if stimuli are sent to an area involved in hunger regulation, animals will eat much more than they normally do. The most starting effects, however, came from the accidental discovery in 1953 by James Olds, then at McGill University in Montreal, that there are pleasure-reward areas near the hypothalamus. Rats allowed to self-stimulate that area preferred the stimulations to food, and some starved to death as a result. In fact, rats self-delivered thousands of stimulations per hour, continuing on for more than a day and stopping only from exhaustion; after a brief rest they resumed the self-stimulation. Since there also are pain-punishment regions near the pleasure-reward areas, scientists have found ESB a powerful technique for administering reward or punishment in conditioning animals to learn skills faster.

Another effect of ESB is to provoke rage or aggression. Stimulating certain parts of the hypothalamus or amygdala of a cat produces hostile and violent behavior. By alternately stimulating adjacent regions of the amygdala, scientists can make animals alternate between outbursts of hate and affection. And when the amygdala is removed from a lynx or killer rat, the animals become tame. As we will see later, this approach also has been tried with people.[11]

By inserting electrodes in different areas of the brain and stimulating each in turn, scientists can use ESB to locate brain tissue associated with such unwanted effects as seizures. This approach has been used with Parkinson's disease, a nervous disorder where the patient has uncontrolled tremors of the hand or foot that slowly spread to other parts of his body. A surgeon can introduce an electrode into a particular region of the thalamus and move it precisely, measuring the electrical activity at each location. When he observes the telltale pattern, he destroys that small area of the brain by heating the electrode. In this way ESB helps him do brain surgery with greater precision.

ESB and microsurgery may also help relieve chronic pain. One experimental approach is to stimulate the brain very gently through electrodes implanted in the pain areas. Another method, which has provided some relief from the pain of terminal cancer, is to destroy certain small clusters of brain cells.[12]

A dramatic use of ESB is to implant electrodes permanently that interact with the brain's own electrical signals. Irving Cooper of Saint Barnabas Hospital in New York, working in collaboration with Roger Avery, a Long Island engineer, developed a brain pacemaker in which electrodes implanted in the cerebellum are connected by wire to a radio transmitter that sends low-voltage stimulation. In theory, the pacemaker generates electric impulses that are strong enough to overcome those produced by damaged brain tissue. In experiments beginning in the early 1970s with people who have cerebral palsy, a paralytic condition where there is permanent brain damage, the pacemaker has helped some patients make remarkable progress in their physical abilities—even to the point of being able to walk.[13] It has also shown promise in treating people with epilepsy.

In short, ESB is the most precise tool we have to map the brain and to locate and destroy offending tissue. Still, there is not a long line of people demanding to have holes drilled in their skulls and electrodes implanted in their brains. For one thing, the risks of ESB far outweigh the benefits for most people, especially when we consider how little we really know about how the brain works. And there is another concern: ESB could be used, at least in theory, to control the way people behave. We will consider that issue later, but now let us move on to the most popular way of altering our brains—drugs.

Drugs

People have used mind-altering drugs for a long time, but we are only beginning to understand how they work. The picture is still cloudy, but scientists are slowly uncovering the secrets of how neurons, singly and in combination, normally work.

Neurons typically have a long fiber at one end (the axon) and hundreds of branching, antennalike fibers at the other end, called dendrites. The dendrites reach out to nearby neurons, establishing communication between a neuron and its neighbors. A nerve impulse traveling along the axon triggers the release of a transmitter substance, which crosses a minute gap (synapse) and binds to the dendrite of an adjacent neuron in such a way that the impulse can cross the gap. Depending on the transmitter, the receiving cell either is stimulated to send the impulse on to another neuron or it is inhibited from doing so. Since each of the ten billion or so neurons in the brain can be powerfully influenced by its neighbors, and both excitation and inhibition signals are possible, efforts to work out the circuitry and communication patterns of the brain should keep neuroscientists off the streets at night for a few more years.

Transmitters play a crucial role in mental processes, and they produce striking effects in animal experiments. For example, when scientists inject noradrenalin (a probable transmitter) into a rat brain in a region just above the hypothalamus, the rat suddenly gets hungry. When they do the same thing with acetylcholine (another likely transmitter), the rat becomes thirsty. But when they inject acetylcholine into a similar region of a cat brain, the cat shows anger or fear, or goes into a trance.[14]

Since some mind-altering drugs chemically resemble transmitters, a popular theory is that those drugs enhance, inhibit, or substitute for the normal action of certain transmitters; drugs that increase the concentration of excitatory transmitters in neurons will raise the level of brain activity, whereas drugs that lower the level of such transmitters will act as depressants. But the evidence for this theory, at least in certain details, is far from conclusive.

Despite our ignorance of how they work, and the enormous problems of abuse, mind-altering drugs are a valuable tool in treating people with mental disorders. In the last three decades, tranquilizers and antidepressants have revolutionized the treatment of manic-depressive and schizophrenic patients, decreasing our use of lobotomies, shock treatments, and straitjackets. Nevertheless, our therapy is far from satisfactory, and we continue to seek more effective drugs.

One promising candidate is lithium carbonate. In clinical tests, lithium compounds have protected people against depression, calmed hyperactive children, and controlled the outbursts of excitement in schizophrenic and, particularly, manic-depressive patients.[15] Despite its promise, though, lithium carbonate is potentially toxic, its side effects have not been studied adequately, and no one knows exactly how it works.

Perhaps the most exciting prospects come from the discovery in 1975 that our brains make several substances, called endorphins and enkephalins, that have morphinelike effects. Studies of their effects are in their infancy, but already these natural opiates have been implicated in pain relief, placebo effects (pain relief from taking inert substances), analgesic effects of acupuncture, narcotics addiction, seizures, and memory.[16] In addition, they may have a powerful effect on behavior. In preliminary studies one compound (beta-endorphin) temporarily relieved symptoms of schizophrenia or depression in several patients. According to Nathan Kline, psychiatrist and director of Rockland Research Institute (Orangeburg, New York):

We realize our experiment was outlandishly expensive. . . . At ten milligrams, our therapeutic dose cost a mere $3000 an injection! . . . But think of what it may mean when we get costs down! Here is this fascinating substance, made right in our bodies, and we found that it worked in schizophrenia, depression, and agoraphobia [fear of unfamiliar environments], as well as addictions. So the conclusion is inevitable: If one substance works in all these situations, maybe they are all based on one biochemical problem. Why, we've probably set psychiatry back 100 years.[17]

Endorphins may be the long-sought link between chemistry and certain mental disorders. Researchers have long believed there is a chemical basis for such disorders, and they base that belief on findings such as: Specific drugs help relieve certain disorders: heredity is a significant factor is such disorders as epilepsy, schizophrenia, and manic-psychosis;[18] dialysis of the blood relieves symptoms of schizophrenia in some patients;[19] and there is some evidence for a correlation between abnormally shaped neurons and mental retardation.[20] If scientists can indeed identify a chemical cause for specific mental illnesses, they may well be able to develop more effective drugs.

In the meantime, our mind-altering drugs affect mostly our general level of alertness. They often produce side effects, and their specific effects depend on us—our expectations, our experience with the drug, and our surroundings. But these limitations, and others, have not kept us from looking inside bottles to find the solutions to our problems. Our mind-altering drugs —stimulants, depressants, antidepressants, hallucinogens, and the like—are grossly overused and misused. Although some may open our minds to new experiences and thus be useful in psychiatric therapy, they are a woefully inadequate solution in themselves.

Biofeedback

For many centuries yogis have claimed they can achieve various mental states at will and mentally control their own oxygen consumption, blood pressure, and pulse. The Western world has long regarded those body functions as "involuntary," so we have tended to dismiss their claims as tricks.

No longer. Our attitude has changed, and for a rather compelling reason: In carefully designed experiments yogis have proved their claims. For example, when H. H. Swami Rama was tested at the Menninger Foundation in the late 1960s, he made one-half of his right palm 10° warmer than the other half. And when electrodes were put on his scalp, he predicted and achieved several different patterns of brain-wave activity.[21] In another experiment, conducted for the British Broadcasting Company in 1970, Ramanand Yogi meditated for nearly six hours inside a locked, air-tight box while scientists monitored his oxygen consumption. They had previously determined his basal rate, which is considered the minimum level for sustaining life in an awake, resting state. While Ramanand meditated, his average oxygen consumption was only 50 percent of his basal rate, and it dropped to 25 percent of his basal rate during one hour of the experiment.[22]

Neal Miller, a psychologist at Rockefeller University, showed that even animals can mentally control "involuntary" functions. Using conditioning techniques such as stimulating the pleasure-reward areas of the brain, Miller's group trained animals to produce specific brain-wave patterns, selectively dilate blood vessels, and control their heart rate, blood pressure, rate of urine formation, and rate of intestinal contractions.[23]

Now it is clear we can mentally control far more of our internal body functions than we had thought. But while the yogis learn to do this through meditation rituals, Miller's work showed a faster way—conditioning techniques. This has led to biofeedback, a method whereby machines help train a person by giving him immediate and continuous information on his progress. For example, a machine monitoring a person's blood pressure might give off a tone or light when the blood pressure is below a certain level.

This instant feedback helps him work out an effective way of mentally achieving his goal.

Biofeedback is not a novel type of learning; we all use feedback—hearing, seeing, feeling, the reactions of others—when we learn. Just imagine how difficult it would be to learn to type, play the piano, ride a bicycle, drive a car, or make a stained-glass window without those forms of feedback. The unique attraction of biofeedback, however, is that we can explore dimensions of ourselves we had thought were inaccessible.

Many people are using biofeedback to relax. By means of headsets that monitor brain-wave patterns, they can learn how to achieve mental states at will. There are four basic categories. The most active state, beta (14–30 hertz [Hz], or cycles per second), occurs when a person is anxious, afraid, or is consciously concentrating on some task. In alpha (8–13 Hz), a tranquil, awake state, he feels serene, relaxed, and receptive to ideas. Theta (4–7 Hz) brings deep reverie and a feeling of the subconscious mind welling up, while delta (0.5–3.5) is deep, dreamless sleep. So by learning how to leave beta and spend a few minutes in alpha, a person can take a pleasant, refreshing break that helps him cope with the rest of a busy day. Relaxation may also prove an alternative to drugs in treating ulcers and insomnia. And biofeedback is being tested as a way to help people lower their blood pressure and pulse, and relieve tension or migraine headaches.

Other uses of biofeedback are only beginning to be explored. For example, spending time in theta or low alpha states might facilitate certain types of problem solving—such as the flashes of creative insight that suddenly and unexpectedly come to people who have been struggling with a problem for a long time; increased time in delta may decrease the time needed for sleep; theta may be useful in psychoanalysis by releasing repressed feelings and experiences; in alpha, the mind may be highly receptive to learning.

The story of biofeedback is just beginning, and it offers some truly exciting prospects. But it is hard to sort out the legitimate progress from the exaggerated claims of charlatans who are capitalizing on the impressive possibilities—and our ignorance. We simply do not have enough data yet to know how effective biofeedback therapy will be. And even where it is effective, people will either have to keep using the machines (which could present logistic and financial problems) or learn how to maintain control without machines. So while biofeedback is a shortcut to yoga in the sense that it helps people develop those abilities faster, we cannot be sure how long they will retain their new powers outside the laboratory unless, like the yogis, they practice regularly and maintain a high level of self-discipline.

Nevertheless, biofeedback is becoming an increasingly useful tool for exploring our minds and achieving new levels of self-control. In fact Gardner Murphy, a psychologist and director of the Menninger Foundation for seventeen years, has optimistically predicted that while some of the claims of biofeedback researchers may turn out to be too bold, most of them will prove not bold enough.[24]

MEMORY

We have seen how brain surgery, ESB, drugs, and biofeedback can alter our physical and emotional activities. But they also affect an even more remarkable quality of our brains—our ability for conscious thinking and learning. Now let us examine that ability, beginning with some explorations into the physical and chemical basis of memory.

Somehow, our brains are able to receive information, select what needs to be saved, store it, and then retrieve it. We do not really know how we do all that, and do it so quickly, but surgery and ESB have provided a few clues. For example, Penfield discovered that when he stimulated certain regions of the cortex, some of his patients vividly recalled a specific experience such as an old tune, giving birth to a baby, or an event from childhood. Yet surgery also has revealed that memory is spread throughout the cortex, and perhaps elsewhere; for operations that remove part of the cortex appear to reduce a person's general ability to remember, but they don't selectively erase specific memories.

We have also learned that the hippocampus, a region in the middle layer of the brain, plays a role in processing or storing new information. In 1953 an epileptic patient in his late twenties underwent an experimental lobotomy procedure that damaged his hippocampi. The operation relieved his seizures, but it caused a new problem: He could remember his experiences prior to 1953, but he could not learn or remember anything after the operations. He lived with his mother but could not get to know new people well. He read the same articles and performed the same puzzles over and over with the same fresh enjoyment and challenge. He could not keep track of where he left objects and, because he moved shortly after the operation, he got lost whenever he went out by himself.

Known simply as H. M., he is famous in the medical literature because his disorder is so clear-cut, and he has been the subject of many studies and interviews. The one, small consolation about his tragic condition is that he did not get tired of all the physicians and journalists asking him the same old questions. But even the interviews posed special difficulties. During one conversation he remarked: "Right now, I'm wondering. Have I done or said anything amiss? You see, at the moment everything looks clear to me, but what happened just before? That's what worries me. It's like waking from a dream; I just don't remember."[25]

Although he could not permanently learn anything new, he did remember some things for a few minutes. This suggests there may be different forms of memory. Indeed, animal experiments point to the same conclusion, so many researchers now carefully distinguish between short-term and long-term memory. An example of the first type is looking up a telephone number and remembering it long enough to place the call. Short-term memories seem to be mostly electric, for they can be disrupted by electric shock, drugs, blows, cold, or other stresses. Long-term memories are not erased this way, so they may involve chemicals.

Are chemicals responsible for specific long-term memories? The debate began in earnest in the late 1950s with the work of James McConnell, a psychologist now at the University of Michigan. McConnell and his coworkers conditioned flatworms to contract when a light flashed; worms that responded at least 92 percent of the time then were cut in half, and each half regenerated a whole flatworm. The regenerated worms—including those that regenerated from tail halves and had to grow new brains—also contracted when a light flashed. From this, McConnell concluded that memory was not stored solely in the brain and it could migrate, so perhaps memories were stored as chemicals.

The next step was to see if "memory" could be transferred to another animal. McConnell's group had untrained flatworms cannibalize worms that had been conditioned to the flash of light. After lunch, those worms learned the response to light much faster than their unenlightened colleagues, whose meal had consisted of untrained brothers.[26]

Suddenly many research groups became interested in memory transfer. Some confirmed McConnell's results; others reported negative results. Soon there were claims of memory transfer in rats, but most such attempts failed. As the pile of conflicting reports grew, so did the suspicion and ill feelings between opposing camps. Those feelings still persist.

If specific memories are really encoded in particular chemicals, and if they can be transferred from one animal to another, it should be possible to isolate and identify some of them. What kind of materials might they be? Experiments in the early 1960s suggested RNA as the "memory molecules,"[27] but later work has implicated proteins. For one thing, RNA helps make proteins in the cell, so the earlier evidence for RNA also applied to proteins. Furthermore, there were studies that pointed directly to proteins. For example, scientists could condition goldfish to swim over a barrier but prevent that learning by administering an antibiotic that blocks the synthesis of proteins but not RNA.[28] Another discovery was that three small proteins are each responsible for transferring a particular "memory" to goldfish.[29]

The most provocative studies linking proteins with memory have come from Georges Ungar and his colleagues at Baylor College of Medicine. They conditioned rats to fear the dark (an unnatural response) and transferred this response by injecting a broth from the rat brains into the abdominal cavities of mice. Next, they prepared a broth from the brains of four thousand trained rats and isolated a specific material that produced fear of the dark when given to untrained rats, mice, goldfish, and cockroaches. They identified the substance, called scotophobin, as a small protein fifteen amino acids long. Then they synthesized scotophobin, injected it into untrained animals, and found it was 67 percent as effective as the natural material in causing fear of the dark. Other researchers have confirmed the effect of scotophobin.[30]

Despite these striking experiments, there is a storm of controversy over the role of small proteins in memory. For one thing, no one knows whether proteins such as scotophobin act solely, specifically, and directly on the memory mechanism. Furthermore, if we needed a different protein for each memory, and if all "memory" proteins were synthesized by the cell's usual method, our genes would have to be carrying the codes for every item and experience we might ever went to remember. That seems unlikely. Nevertheless, Ungar envisions a chemical code where certain amino acid sequences (proteins) are associated with certain memories. Even though the number of possible memories we might store in our lifetime is enormous, so is the number of different proteins containing a sequence of, for example, fifteen amino acids. (That particular number is about 10,000,000,000,000,000,000.) Perhaps the proteins would serve as switches to establish and reinforce circuits in the brain for recalling specific experiences.

Nobody has much idea of how it might work, but Karl Pribram of Stanford University has suggested our brains might somehow store memories like a hologram.[31] A hologram is a piece of film that bears the image of a scene photographed in a special way with laser light. The hologram is meaningless in normal light, but when laser light shines through, the scene appears in three dimensions. And even if only a small corner of the hologram is used, the entire scene still appears. It is something like the ability of a simple object or other stimulus to stir our memory of a more detailed picture, or even an entire sequence of events.

If memory is a chemical process, at least in part, we should be able to improve or preserve memory with drugs. After the first reports of memory transfer, there were lighthearted (and lightheaded) speculations that injecting the right chemicals, perhaps extracted from wise old professors, would be an efficient way of giving knowledge to others.

Animal experiments suggest less spectacular, but more realistic, possibilities. For example, conditioned mice whose pituitary glands are removed cannot remember certain old tasks or relearn them; but those abilities return when they receive three pituitary hormones (ACTH, MSH, and vasopressin). And rats and mice can be conditioned to learn more quickly when they receive drugs such as strychnine, amphetamine, nicotine, and picrotoxin. There is an upper limit on the doses, though, both in terms of the effectiveness of learning and because of the experimental difficulties in measuring posthumous learning.[32]

Attempts to improve human memory with drugs have generally failed. In the 1960s, for example, a scientist reported that magnesium pemoline (Cylert) improved the memories of senile men, but his results were not confirmed in other studies.[33] Nevertheless, the search goes on. One candidate, a drug called piracetam, reportedly increased the learning ability of university students and is being tested for its effectiveness against senility and mental retardation.[34] And in 1978 three drugs (physostigmine, arecholine, and choline chloride) that are thought to increase the concentration of acetylcholine in the brain were reported to improve human long-term memory as measured by word-recall tests.[35]

Pharmaceutical companies are interested in developing a drug that improves memory and learning, for there would be an enormous market for such a product. For example, we could require our elected officials to take the drug every two months in order to recall their campaign promises. But the largest market would be the prematurely senile and the elderly, who have lost millions of neurons. Because they have fewer neurons, and the ones they still have may not work as well as they used to, the elderly may find it hard to store information, or retrieve it, or both. While such problems will be very difficult to solve, there is no shortage of optimists. Indeed, Ewald Busse of Duke University predicted (in 1972) that in ten years we would have a drug or other treatment to retard (if that is the right word) the aging process in our brains.[36]

Don't hold your breath waiting for it to happen. For one thing, our ignorance about learning and memory is enormous. Moreover, our feelings play an important role in memory, for we are more likely to have our emotional experiences emblazoned on our minds. Another complication is "state-dependent" learning, a phenomenon where something we learn under the influence of certain drugs is difficult to recall except when we are again under the influence of that drug. How often we use certain information also affects how easily we can recall it. Which explains, for example, why we forget so much of what we see on television.

INTELLIGENCE

It is our intelligence—our ability to learn, to think rationally and abstractly, and to apply our knowledge—that sets us apart from the rest of creation. It has helped us gain vast control over our physical and social environments, and it underlies our constant quest for a better life. Yet the level of intelligence we now possess may be only the beginning, and we may be able to use it to improve on nature's design. Because evolution is so slow, and intelligent beings have been on earth a relatively short time, Robert Sinsheimer has speculated that there is considerable room for improvement. Of our current, modest status he has asked: "What else can we sensibly expect, when we are apparently the first creatures with any significant capacity for abstract thought?"[37]

How could we increase man's intelligence? One idea is to increase the size of his brain (though the dividends in intelligence might be small because even now we use only a fraction of our potential). Man's brain reached its current size at least 40,000 years ago, and some people 250,000 years ago had brains as large as ours. Perhaps genetic manipulations would work, or we could find a way to coax the brain cells into dividing one more time, thus doubling the size of the brain. Still, we must recognize a built-in limit with the way we reproduce: the fetal brain must be small enough to squeeze through the birth canal. So any method that significantly increased the size of fetal brains would have to be accompanied by the routine use of cesarean sections or artificial wombs.

Since the genetic engineering of humans lies far in the future, our first attempts to increase intelligence will be in helping each person reach his genetic potential. Here the most effective therapy will occur early in life, when the brain's performance is developing the most rapidly.

One of our most effective tools is good nutrition. In experiments where pregnant rats ate food deficient in calories, protein, or essential amino acids, their babies had low body weights and their brains were low in protein, number of cells, and total weight.[38] And rats from underfed mothers seem to show such behavioral disorders as a poor ability to learn mazes.[39] Malnutrition also harms people. A major health problem in regions of Asia, Africa, and Latin America is kwashiorkor, a condition of protein deficiency in early childhood. Numerous studies there show malnutrition during the early months of life is associated with permanent intellectual damage.[40] Indeed, of all our efforts to increase man's intelligence, the greatest improvement by far, worldwide, could come from providing proper nutrition in the early years of life.

Another possibility is to use hormones to increase brain development early in life. In one study rat fetuses receiving growth hormone reportedly developed larger brains and better learning ability.[41] In other studies growth hormone eliminated much of the damage to offspring born from rats on calorie-deficient diets.[42] But whether this approach would be safe, practical, and effective with human pregnancies is anybody's guess.

O. S. Heyns, a South African physician, reported in the late 1960s that unusually bright babies were born when he increased the oxygen supply to fetuses during the ten days or so before delivery and during birth. He did this by having the pregnant women use an abdominal decompression chamber, which reduced the pressure of the abdominal muscles on the uterus. According to Heyns, fetal brains may not fully develop during the last days of pregnancy because they need more oxygen than the mother normally provides.[43] This claim has not been established, nor has the effectiveness of his treatment, but there is no doubt that oxygen deprivation prior to, or during, birth can permanently damage the brain.

External stimuli are yet another way to influence brain development. For example, if a kitten's eye stays closed for a few weeks, beginning about the fourth week after birth, the permanent result will be that fewer of its brain cells can react to signals from that eye.[44] Furthermore, a kitten cannot recognize horizontal and vertical lines unless it sees them early in life.[45] A more general effect of environment was shown in experiments where rats lived in uninteresting surroundings while others roamed a large, object-filled area. The brains of the latter group had heavier and thicker cortexes, more glial cells, larger neurons, more interconnections between neurons, and larger capillary diameters. And those animals initially (but not permanently) showed a better ability to solve problems.[46] And in yet another study baby rats placed in an enriched environment for four weeks had increased branching in the dendrites of their hippocampus cells, but the enriched environment had no such effect on adult rats.[47] Although experiments with human babies cannot be as clear-cut, studies of infants who were deprived of normal stimuli, or who received an unusually rich environment, indicate we might be able to increase brain development during the first year or so after birth by using the right stimuli: visual patterns (objects with stripes, polka dots, various color combinations, different shapes and sizes), sounds from a radio, moving objects, people talking, rocking motions, handling the baby, and giving him objects to reach for and manipulate.

One of the most startling ideas for increasing our intelligence is to expand the capacity of our brains by integrating them with computers. Indeed, the growing sophistication of brain pacemakers and other devices that send impulses directly to the brain suggest the possibility of implantable computers. According to Adam Reed of Rockefeller University: "As an aid to memory, [they] should provide the user with an almost indefinite data capacity."[48]

Yet even if we had that technology, it is unlikely that someday each person would have a computer implanted in his brain. For one thing, it will simply be too expensive. Furthermore, we can reap the main advantages of computers—storing and rapidly processing enormous bits of data—without implanting them in our heads and compromising our freedom to think independently.

Could we improve on mankind's "intelligence" by building external computers that can think and reason? Computers are already intelligent in the limited sense that they can rationally examine possible courses of action and the likely consequences of each course. For example, computers have been programmed to play complex games such as chess. But first a human being must supply the computer with the necessary information, and the instructions on how to process that information. And although the human brain cannot match the speed of the computer in manipulating large amounts of information, our brains have other characteristics that a computer cannot fully duplicate: the ability to interpret nuances of our language, such as inflection, context, and nonverbal communication; the ability to have, and learn from, human experiences; and the capacity for creative thought, intuition, curiosity, and wonder.

We will make increasingly complex and intelligent computers, but they need not threaten us—provided we realize computers are tools for exploring options, not for making actual decisions. We must recognize and evaluate the assumptions that are built into each computer program, for if the input is faulty, so is the output (garbage in: garbage out). Furthermore, a computer cannot incorporate feelings into the decision-making process. In a humane society, however, we must temper our decisions with a sense of compassion and respect for others, and for ourselves. As Lewis Thomas has noted in *The Lives of a Cell*: "The future is too interesting and dangerous to be entrusted to any predictable, reliable agency. We need all the fallibility we can get. Most of all, we need to preserve the absolute unpredictability and total improbability of our connected minds. That way we can keep open all the options, as we have in the past."⁴⁹

SOME CONCERNS

Restrictions on Brain Experimentation

Surgery, ESB, and mind-altering drugs are two-edged swords. They may help people with intense and chronic pain, senility, epilepsy, cerebral palsy, schizophrenia, or manic depression; they also could rob people of their full humanity. When we use these techniques, the stakes are very high.

When is it right to use experimental methods on human beings? A standard answer is that adults may participate in medical experiments if they give their voluntary and informed consent in advance. But that is not much help to people with brain disorders, many of whom are incompetent to make reasonable decisions for themselves and must let someone else decide for them.

Psychosurgery treats mental disorders, so these operations have been performed on relatively defenseless people, many of them in mental institutions. Prisoners have been another convenient source of subjects. Members of some minority groups also say they too often have been selected for psychosurgery.⁵⁰ And children can claim a similar distinction. For example, psychosurgery has been used in Japan to calm hyperactive children.⁵¹ A

Mississippi physician, who has done psychosurgery on children—most of them in institutions—has explained that the operations are necessary "for custodial purposes, when patients require inordinate amounts of care."[52]

During the 1950s and 1960s scientists did a variety of ESB experiments on human subjects. In one study, conducted over a seven-year period, Robert Heath and Walter Mickle of Tulane University School of Medicine implanted electrodes in the brains of fifty-two patients, mostly chronic schizophrenics. Depending on which area was stimulated, they reported five types of response: good feeling, alertness, and relief from pain; discomfort and anxiety; tension and rage; drowsiness; and fear or rage. In another study, patients at the Gauster Mental Hospital in Oslo, Norway, were allowed to control their own ESB buttons; they learned to avoid negative regions, and they stimulated their pleasure areas over and over, sometimes to the point of convulsions.[53]

Can people adequately protect themselves against misuses of mind-altering techniques? Those who cannot give their informed consent must rely entirely on the judgment of others. Indeed, all of us depend on a traditional safeguard against dangerous treatment—the professional judgment of the physician. Yet the excessive prescriptions for stimulants, tranquilizers, antidepressants, and the like suggest that too many doctors succumb to our demands for the magic potions we think will chase away our problems. And while few physicians are qualified to operate on the brain, even at this level there is not a consistent policy on treatment, for there are honest differences of opinion as to what is appropriate. So where does that leave us? According to Herbert Vaughan, a neurologist at the Einstein College of Medicine: "There are no legal controls on psychosurgery, no safeguards at the present time except the conscience of the physician. . . . It is a strictly *laissez faire*, entrepreneurial relationship."[54]

We need medical guidelines that specify when mind-altering techniques may be used. This is especially important for psychosurgery and ESB, which permanently alter the brain. Medical associations, together with people outside medicine, have developed guidelines for other difficult matters (diagnosing death; the continued use of life-sustaining machines), and they could do the same for brain-altering techniques. In addition to medical criteria, we need committees with no vested interest but with the competence and authority to approve psychosurgery in advance and to ensure a thorough followup on the patients. The latter is a glaring deficiency of our current nonsystem. According to Ernest Bates, a neurosurgeon at the University of California Medical Center, San Francisco: "Psychosurgeons should be trying to learn all they can about the normal functions of the structures they remove or injure. This can be done only through testing of all patients who have had such operations. Human nature is such that most surgeons probably do not want to acknowledge deficit in their patients even though such findings would advance our knowledge of brain function and help refine psychosurgical methods."[55]

For now, there is no central registry for psychosurgery, no assurance the same criteria are used by all surgeons, and the extent of patient follow-up is left to the discretion of each physician. That is not good enough.

Controlling Behavior

We have many ways to control the behavior of others—brute force, threats, advertisements, laws, political propaganda, social approval or disapproval. And now our advancing technology suggests another tool: direct manipulation of the brain.

We might imagine, for example, a ruthless dictator who uses ESB by remote control to keep his subjects passive; at the same time, he commands a loyal force of soldiers who become aggressive whenever he presses the right button. Indeed, Delgado has shown that ESB can produce passice or aggressive behavior in animals, and its effects can apply to rulers as well as their subjects. In one experiment, for example, a monkey dictator with implanted electrodes became apathetic whenever he was stimulated by the push of a lever. When the other monkeys in his colony discovered the lever and its effect, their ruler experienced a unique power failure.[56]

Nevertheless, the idea of a ruler controlling his subjects by ESB or psychosurgery is strictly science fiction. For one thing, individual people (and animals) vary considerably in their brain structure, and in their responses to the same manipulation. Their specific responses depend not only on the manipulation itself, but also on the environment, their experiences, and their own physical and personal qualities. So even if a dictator could handle the logistics of doing brain operations on large numbers of unwilling people and solve the technical problems (for example, electrodes coming loose, scar tissue forming and producing unexpected behavior), he would still be disappointed with the results. In short, it will continue to be far more efficient for a tyrant to stay in power by using political propaganda, threats, clubs, and guns (and not necessarily in that order). As Joshua Lederberg has remarked: "It is indeed true that I might fear the control of my behavior through electrical imulses directed into my brain but . . . I do not accept the implantation of the electrodes except at the point of a gun: the gun is the problem."[57]

A greater problem is deciding when we should use these methods to try controlling human behavior. Beginning in the 1960s psychosurgery has been used therapeutically on people who consistently show uncontrollable and violent behavior. Keiji Sano, a neurosurgeon at the University of Tokyo, reported that forty-six of fifty-six patients (many of them children) with intractable epilepsy plus mental retardation, hyperactivity, and violent behavior became more tranquil after psychosurgery.[58] And in the United

States the leading practitioners are Vernon Mark, William Sweet, and Frank Ervin (all associated with the Harvard Medical School until the early 1970s, when Ervin moved to UCLA). Over a seven-year period, they operated on thirteen people whose violent behavior was related to a rare type of epilepsy; several of their patients improved after all or part of their amygdalas were destroyed. In their controversial book *Violence and the Brain* Mark and Ervin endorsed psychosurgery as a way to treat certain types of people who have uncontrolled outbursts of violence and rage. They've also suggested it might be possible to analyze people's brains so as to identify in advance individuals who are prone to violence.[59]

Where do we draw the line in using ESB and psychosurgery to curb unacceptable behavior? The answer is not to ban them completely, for someday they may help people with mental disorders return to the mainstream of human life. But we need medical guidelines specifying that we will use these techniques only when drugs and other types of therapy have failed, the brain is seriously abnormal (especially where there is organic damage), and there is a reasonable basis for thinking the operation might help. Otherwise, we might be tempted to label prisoners and other nonconformists as mentally ill and try to force them into "acceptable" patterns of behavior by operating on their brains.

Would we ever try to control behavior to that extent? We certainly are not shy about using other methods, such as conditioning techniques, to instill specific values in our children and teach them to behave in socially acceptable ways. John B. Watson, one of the founders of behavioral psychology, said in 1925: "Give me a dozen healthy infants, well formed, and my own special world to bring them up in, and I'll guarantee to take any one at random and train him to become any type of specialist I might select—doctor, lawyer, artist, merchant-chief and, yes, even beggarman and thief, regardless of his talents, penchants, tendencies, abilities, vocation, and race of his ancestry."[60]

The crucial question, however, is not whether we can control people's behavior—the answer to that question clearly is yes—but whether it is right to do so, whatever the method. According to Herbert Kelman, a social psychologist at Harvard University: "For those of us who hold the enhancement of man's freedom of choice as a fundamental value, any manipulation of the behavior of others constitutes a violation of their essential humanity, regardless of the 'goodness' of the cause that this manipulation is designed to serve."[61]

If we are to progress as a humane society, we need the rebels to aggravate us, question our motives, and dispute our assumptions and judgments. Even more important, though, as human beings we each deserve the freedom to think and act within the broadest limits society can tolerate. That may make it harder to maintain order. But a highly ordered society is a poor bargain if the price of admission is our humanity.

Self-Identity

Thomas Edison said: "The body is just something to carry the brain around in."[62] We know that is too simple, for we are coming to realize the exquisite interdependence of the brain and the rest of the body—how emotional, chemical, and physical events in the brain influence the rest of the body, and how experiences of the body affect the brain. Nevertheless, we also know the brain is central to our identity as human beings.

Our growing technology for manipulating the brain has the potential both to enhance and to threaten our status as free, rational, emotional, and moral beings. But there may be limits to how far we can go. Indeed, it is not clear, even in principle, that man's mind is capable of fully understanding man's mind. Although we will greatly expand our knowledge of the human brain, we will probably never know, in physical and chemical detail, exactly why we behave as we do. For one thing, each of us has a unique set of experiences that affects our behavior. Furthermore, we are emotional beings, and a complete understanding of feelings may be unattainable by the rational methods of science. And while scientists can provoke a few emotional responses by physically and chemically manipulating the brain, that is not the same as fully understanding our reactions to events around us, how they were shaped by past experiences, and how they will affect our actions in the future.

Western tradition commonly has held that reason is good and emotion is not so good. We have the image of the modern technocrat—rational, efficient, apparently unemotional, and immune from having his sense of reason compromised by his feelings. Yet feelings are an important part of our identity as fully human beings, and we should not ignore or deny them. Indeed, someone once said, "I don't trust a man who never cries." We need a proper balance of reason and emotion, for life without emotion is meaningless.

Some people may regard mind-manipulating techniques as a way to discover themselves and find meaning in life. The most popular method is drugs, but more sophisticated techniques are on the way. According to a survey of psychology experts, do-it-yourself pleasure centers will be possible about the turn of this century.[63] Herman Kahn of the Hudson Institute has envisioned people having chest consoles with various wires connected to their pleasure areas, and he fantasizes: "Any two consenting adults might play their consoles together. Just imagine all the possible combinations. . . . But I don't think you should play your own console; that would be depraved."[64]

Our leisure activities will become increasingly elegant as we seek external ways to inject novelty, excitement, and satisfaction into our lives. We will

devise more specific ways to manipulate our minds. We will discover exotic
new forms of entertainment. But these diversions, in themselves, will not
bring us deep and lasting happiness, for that must come from within. There is
no adequate substitute for the experience of personal exploration, discovery,
and growth. As Richard Feynman, a physicist at the California Institute of
Technology and a Nobel laureate, explained: "My father got me interested in
all these things by telling me how wonderful nature was. But he couldn't
know the terrible excitement of making a new discovery. You get so excited
you can't calculate, you can't think any more. It isn't just that nature's wonderful because if someone tells me the answer to a problem I'm working on,
it's nowhere near as exciting as if I work it out myself."[65]

That is what gives life so much of its savor—the chance and challenge to do
it ourselves, with our own abilities, striving for our best. When we consider
techniques for modifying the brain, we must judge whether they help a person to discover and develop his own ideas, interests, and abilities, and forge
positive, genuine relationships with others. For that is the voyage that makes
human life worthwhile.

The coming generation will be the last generation to seize control over technology before technology has irreversibly seized control over it. A generation is not much time, but it is some time.

Robert L. Heilbroner

Of all things, people are the most precious.

Mao Tse-tung

9
Science and Society

As scientists boldly uncover the secrets of nature, they pull society along in their wake. We have eagerly accepted the fruits of that partnership—victory over infectious diseases, pest control, better communication devices, new amusements, convenience foods, more leisure time, and longer, healthier lives. Yet it has not been an entirely happy marriage, for we hold science (and its companion, technology) at least partly responsible for such delights as nuclear weapons, polluted air and water, overpopulation, and our increasing exposure to hazardous chemicals. And our growing control over the mechanics of life brings a more subtle threat: turning people into things. So it is important for us to examine the interplay between society and science, and see how we can reap the benefits of scientific knowledge while keeping our losses at a minimum.

Let us begin with a specific example. When scientists discovered how to splice foreign DNA into plasmids, they suddenly opened the door to the genetic engineering of bacteria. They also ignited an international debate about the conduct of scientists in society, a debate that is still simmering. While the immediate issue is recombinant DNA research, the broader issue is the freedom of scientists to pursue knowledge, and the rights of society to restrict them.

THE RECOMBINANT DNA DEBATE

In chapter 6 we examined what the scientists have done to reduce the risks from gene transfer experiments. First, they called for a moratorium on the most dangerous experiments, and they observed that moratorium until the conference at Asilomar adopted safety guidelines. Although during World

135

War II physicists had asked their colleagues to stop publishing atomic data, never before had scientists called a halt to their own research. Then NIH adopted safety rules that apply to all institutions receiving grants from that agency. And many European nations have adopted either those rules or the British rules, which put more emphasis on physical, rather than biological, containment.

In the United States, however, the current arrangement has provoked a storm of criticism. A few critics argue that the rules are not strict enough. For example, they point out that *E. coli* was chosen as the host organism for reasons of convenience rather than safety; namely, *E. coli* is the best-studied microbe and it would take years to develop another candidate. And even though weakened strains are used, they believe it would be safer to find a different type of microbe, one that does not naturally live inside people. The critics also contend that the rules, however stringent in themselves, cannot adequately protect against human error. And even a very small risk of an accident, they say, is unacceptably high in view of the possible damage. George Wald, a Nobel laureate at Harvard University, said: "Recombinant DNA faces our society with problems unprecedented not only in the history of science but of life on the Earth. It places in human hands the capacity to redesign living organisms. . . . The results will be new organisms, self perpetuating and, hence, permanent. Once created, they cannot be recalled."[1]

Most scientists, however, think the NIH rules provide an ample margin of safety. In fact, many consider the restrictions excessive, thus making recombinant DNA research needlessly slow and expensive. They point to the excellent safety record where similar containment methods were used on the most virulent microbes known to man. Gene splicing with those organisms is banned, and scientists do not actually know whether their experiments would create microbes as hazardous as the ones we already have. Besides, bacteria naturally exchange bits of genetic information by plasmids, so the idea of scientists making microbes with new combinations of genes is not necessarily ominous. Lynn Elwell of the Burroughs-Wellcome Company and Stanley Falkow of the University of Washington have written: ". . . while committees of scientists and laymen banter about recombinant DNA around the conference table, nature . . . has been conducting experiments prohibited under the NIH guidelines for recombinant DNA research."[2]

The best evidence of safety will come from the experiments themselves. The record so far is good; in fact, there have been no cases of infection reported during several years of recombinant DNA research. One reason, John Abelson of the University of California, San Diego, has suggested, is that bacteria are unlikely to accept virulent genes and survive. "Thus," he contends, "it is probably not possible to create a strain that would overgrow the laboratory and head for the town as depicted in movies of the 1950s."[3]

Nevertheless, scientists want to test certain "worst case" situations, which require the most elaborate (P4) facilities. They plan to splice into bacterial plasmids the DNA from a virus that causes tumors in mice (but not in humans). Then mice will eat bacteria, breathe them, and get injections of them while scientists monitor the animals to see if they develop tumors. The first results, reported in 1979, show mice are not infected when they are fed or injected with those bacteria, but injections of the recombinant DNA itself may cause infections.

Whether the safety rules are strict enough is not the only concern. Critics say the current arrangement has too many loopholes, for the NIH rules do not have to be followed by scientists doing military, industrial, or other private research. And recombinant DNA techniques are simple enough that, according to Paul Berg, "some of these experiments can probably be done as a high-school exercise."[4]

Enforcement is another issue. Critics argue that since NIH is composed mostly of scientists and its major function is to promote research, there is a conflict of interest in having that agency serve as the watchdog of scientific research. Of this self-policing arrangement, one scientist remarked: "This is probably the first time in history that the incendiaries formed their own fire brigade."[5]

Indeed, the NIH rules give scientists considerable latitude in deciding how rigorously to monitor their research. And there is already evidence that the current arrangement works poorly. For there have been two suspected violations of the rules—one was a real violation and the other was confusion over red tape—and in both cases there was a lag of several weeks before NIH even heard of the problem.

Another criticism is that the scientists themselves drew up the safety rules. Speaking of the original moratorium and the conference that adopted provisional guidelines, Senator Edward M. Kennedy said:

For literally the first time in the history of science, researchers stopped their work to consider its implications; to see if they should go forward at all. . . . It was commendable, but it was inadequate. It was inadequate because scientists alone decided to impose the moratorium and scientists alone decided to lift it. Yet the factors under consideration extend far beyond their technical competence. In fact they were making public policy. And they were making it in private.[6]

It has indeed become a public issue. The first official action came in 1976, when the Cambridge City Council, having learned of Harvard's plan to build a P3 facility, held hearings on whether such research would endanger the citizens of Cambridge. After hearing Nobel laureates and others disagree with each other, the council voted for a three-month moratorium on P3-level research and set up an advisory board of citizens to study the matter further.

Seven months later, after holding extensive hearings of their own, the citizens' advisory group recommended that P3 research be allowed in Cambridge under the NIH rules but with a safety committee of local citizens to monitor the work. The council accepted those recommendations and the ban was lifted.

It was just the beginning. Bills to regulate recombinant DNA research have been introduced in several cities and states, and in 1977 the Maryland General Assembly passed a law extending the NIH guidelines to private companies. The most likely outcome, however, is federal legislation. One reason is that redesigned microbes are illiterate; they tend to ignore our laws prohibiting them from crossing city or state lines. Federal legislation would also prevent a jumble of state and local laws (though possibly allowing communities to adopt stricter rules in the unlikely event they could prove they were necessary).

Several senators and representatives have introduced bills that would require everyone doing recombinant DNA research in the United States to follow the major provisions of the NIH rules. Most bills would provide for the licensing and inspection of research facilities and the power to levy penalties for violations. They also address the special needs of private companies, such as keeping their research plans confidential, doing experiments on a large scale, and protecting proprietary rights. (Indeed, Washington's Court of Customs and Patent Appeals ruled in 1978 that General Electric could patent a new form of life, a microbe they made by transplanting plasmids.[7]) Some regulation could already be done under our existing laws—for example, the right of the Environmental Protection Agency (EPA) to regulate toxic substances, and the authority of the Occupational Safety and Health Administration (OSHA) to protect laboratory workers—but specific federal action is needed to ensure that the safety rules will apply to everyone. Canada has already taken that step. In 1978 the Canadian government announced that it will require all gene-splicing research in Canada to conform to the guidelines of their Medical Research Council (which are similar to the NIH rules).

The fuss over recombinant DNA has brought politicians into the fray, and some of their efforts to regulate and restrict research bother scientists. But there seems no way out. According to David Clem, a Cambridge city councilor who voted for the three-month moratorium; "I tried to understand the science, but I decided I couldn't make a legitimate assessment of the risk. When I realized I couldn't decide to vote for or against a moratorium on scientific grounds, I shifted to the political."[8]

Which brings us to the heart of the matter—the relationship between science and the whole of society, as represented by our political institutions.

Now let us turn to this broader issue.

THE CASE FOR FREE SCIENTIFIC INQUIRY

Some scientists fear the safety issue will be used as a wedge to impose further restrictions on their research. They can offer a stout defense against such efforts, so let us give them the stage for a few minutes.

Their first argument points to the misguided attempts in the past to suppress research. The most famous case is Galileo, who developed the telescope and used it to study the heavens. His observations convinced him that the center of our solar system is not the earth, but the sun. Yet the Church, armed with righteousness and ignorance, rejected this new information and attacked Galileo personally. In the face of threats that he would be physically harmed unless he recanted, Galileo publically confessed his "error." But when the truth finally emerged, it did not damage man's concept of God; instead, it damaged the credibility of the Church. As Cardinal Suenens said at Vatican II, "One Galileo trial is enough for the Church."[9]

Another example is T. D. Lysenko, a biologist who rose to power in the U.S.S.R. He believed organisms could pass their environmentally acquired characteristics on to their offspring, and he crushed scientific opposition to his theories. After decades, Russia is just now recovering from the disastrous effects on its plant and animal breeding programs, and on its basic research in genetics. So the take-home lesson, according to Philip Handler, president of the National Academy of Sciences, is: "The foolish government that knowingly interferes with the course of science, corrupting or perverting its outcome, will itself be the inevitable victim of that crime."[10]

History is not the only argument for free scientific inquiry. Some of the others are in the litany that follows:

"Even if it can be delayed, the emergence of new scientific knowledge cannot be stopped. There is no power on earth that can do that, for the good and simple reason that what is unknown cannot be inhibited."[11]

"Nor can we unlearn the scientific method, which is available for all who wish to wrest secrets from nature."[12]

"Those who consider themselves guardians of the public safety must count the costs to the public health of *impeding* research, as well as the speculative *hazards* of research."[13]

In science we trust.

Knowledge is ethically neutral. Therefore, we should not interfere with scientists who are acquiring basic knowledge, nor hold them responsible for eventual misuses of that knowledge. Only the applications of knowledge require ethical judgments, and it is only here that we should intervene.

> "Knowledge is power, and power can be used for good or for evil; and, since the genie that brings new knowledge is already out of the bottle, we must learn to direct the use of the resulting power rather than curse the genie or try to confine him."[14]

Yet some knowledge carries a clear potential for misuse. Should not the scientist show restraint, or be restrained, in developing new knowledge?

> "Since our capacity for reliable long-range technology assessment is virtually nil, vigilance concerning new knowledge that might someday be misused is a threat to freedom of inquiry, and I believe a threat to human welfare."[15]

Is some knowledge too dangerous to have?

> ". . . it is not dangerous knowledge but dangerous ignorance that we have to fear. . . . The answer to dangerous knowledge continues to be more knowledge, broadly shared by an international community of science and scholarship, and reliance upon the determination of man to grow in wisdom and understanding."[16]

Is there knowledge we are not ready for, knowledge we should not seek until we have the wisdom to use it for the collective benefit of mankind?

> ". . . it is impossible to wait with new explorations until man is a more moral being. It is not likely that he will soon attain the status of an angel, either by changing his basic nature or by being born into a perfect society."[17]

We are by nature curious and creative beings, and our search for knowledge and truth is a mark of our humanity; indeed, it is the key to our further progress as human beings. We cannot change or deny truth because of political, social, religious, scientific, or any other views that happen to be fashionable at the time. The quest must go on, unhindered, whether what we learn is popular or not.

> "The precedent of beginning to censor research because of the knowledge

it might lead to . . . is extremely dangerous. If we want workable solutions to social problems, they have to be built on a reality that's there."[18]

"If you're dedicated to the truth, you have to say that there are no truths not worth seeking."[19]

Amen.

RESTRICTIONS ON SCIENTIFIC INQUIRY

Absolutes have a nice ring and make us want to stand up and cheer. But they tend to pale in the harsh light of reality. There are few absolutes in life, just priorities.

All the arguments for free scientific inquiry have merit, but they come under suspicion, at least when scientists offer them, because they are self-serving. Their practical effect is that society should leave scientists alone to do what they want in their search for basic knowledge. Only when that knowledge is applied should nonscientists raise their voices.

But where does that freedom of inquiry come from? Society. Who supplies the money (directly or indirectly) for scientific research? Society. And where is the impact of new knowledge felt most? Society.

Science is an adventure of the human spirit, no more or less so than music, history, economics, poetry, athletics, art, sociology, and philosophy. There is no divine right for science, or any other human activity, to be exempt from the political process, which operates (at least in theory) for the best interests of society as a whole. Now we may well decide our wisest course is to keep restrictions at a minimum—as in separating Church and State—but that still is a political decision.

Just as we have no absolute freedom to pursue life, liberty, and happiness, scientists have no absolute freedom to pursue knowledge. The purpose of limiting freedom is to expand it for society as a whole. According to Garrett Hardin, a biologist at the University of California, Santa Barbara; "When men mutually agreed to pass laws against robbing, mankind became more free, not less free."[20]

The pursuit of knowledge and understanding is a right we should cherish and protect. But society has the right—indeed, the duty—to limit that freedom when necessary. The practical problem is deciding when it is truly necessary, for history has amply demonstrated that this power should be used sparingly. Nevertheless, we can identify several reasons for restricting scientific inquiry.

Public Safety

Our political institutions have a clear obligation to restrict research that threatens public health. No one, the scientists included, disputes this. Indeed, we have seen how this concern led to the rules on recombinant DNA research. And this is hardly an isolated example, for we also have safety regulations restricting experiments with radioactive, carcinogenic, and other hazardous materials. Here the issue is not *what* may be learned, but *how* it may be learned. The freedom to pursue knowledge is not a license to do so any way the experimenter chooses.

But are political bodies competent to rule on the public health risks of scientific research? Do they have any choice but to leave those decisions to the scientists themselves? Here, again, the controversy over recombinant DNA is instructive. According to an editorial in the *Medical Tribune*: "It is the scientists who have pointed out the hazards; who are, worldwide, considering and introducing the methods to contain them; and who are best equipped to direct the scientific course of the investigations."[21]

But a different answer comes from the Cambridge citizens' advisory group. After hearing expert testimony on the dangers of recombinant DNA research, they decided the NIH rules were basically adequate, but the procedures for monitoring the research were deficient. By their thoughtful deliberations and conclusions, they amply demonstrated that informed citizens are able to make sound decisions. And they affirmed their right to do so, declaring: "Knowledge, whether for its own sake or for its potential benefits to humankind, cannot serve as a justification for introducing risks to the public unless an informed citizenry is willing to accept those risks. Decisions regarding the appropriate course between risks and benefits of potentially dangerous scientific inquiry must not be adjudicated within the inner circles of the scientific establishment."[22]

Resources

Another restriction is money. We simply cannot afford to do everything we would like to do, so we will have to make some hard choices. "There are no truths not worth seeking" may be true philosophically, but not financially.

What will we choose? We have seen there is no shortage of projects to spend our money on: separating X and Y sperm so parents can choose the sex of their child; test-tube fertilization and embryo implantation to help infertile couples have children; cloning livestock to increase our food supply; finding a way to diagnose muscular dystrophy in a fetus; redesigning microbes to produce the clotting factors hemophiliacs need; developing an artificial heart; finding better ways to treat schizophrenia; and many more.

We may well decide to allocate our research funds so as to provide the greatest benefits to the largest number of people. Yet that will not leave much for the scientist who is trying to develop a cure for a fairly rare disease, or a method to diagnose its presence in a fetus. The decisions are daunting, but inescapable.

How can we make wise decisions? The universal solution to knotty political problems is to create an omniscient committee (or "another bureaucracy," depending on your point of view). One plan is to establish a science court, a group of competent and neutral scientists who would analyze the scientific aspects of public issues and make recommendations to the public policy makers. Plan B is to create a national commission, composed of scientists and other leaders, to assess the impact of various areas of research and recommend priorities to the funding agencies. In either case, our "assessors" would have to weigh the costs and benefits of saying yes, and also of saying no, taking into account both the short- and long-term effects. The trouble is that omniscience is in rather short supply these days, and "the relationship between research and prosperity [is] as predictable as that between prayer and deliverance."[23]

While nonscientists should help decide where we spend our research funds, they are generally not competent to judge individual research proposals. Here scientists have staunchly defended a peer review system. For example, according to a survey in 1976 of National Science Foundation (NSF) grant applicants and reviewers, the NSF peer review system is fair, works well, and needs few, if any, changes.[24] At the same time, scientists and their funding agencies are sensitive to public pressure for research promising imminent benefits. So promises abound. Since the key to a problem often comes from an unexpected source, almost any claim can be semijustified. Macfarlane Burnet, an Australian immunologist and Nobel laureate, has described one example as follows:

I think [we should] look rather critically at that perennially repeated justification for work in molecular biology—that all competently done research in fundamental aspects of biology will help toward discovering the cause and cure of cancer. I believe that most scientists who make this claim, usually are virtually compelled by social forces to tell this white lie with as much apparent conviction as they can muster. They know that their own work is rated as good by their peers, who are concerned not at all with its bearing on cancer but deeply with its originality, its integrity of approach and interpretation, the elegance of the methods used, and in implications it will have for the interpretation of other biological phenomena. They are rightly proud of their achievement and equally rightly feel that they have won the right to go on with their researches. But their money comes from politicians, bankers, foundations who are not capable of recognizing the nature of the scientist's attitude to science and who still feel, as I felt myself thirty years ago, that medical research is concerned only in preventing or curing human disease. So our scientists say what is expected of them, their grants are renewed and both sides are uneasily aware that it has all been a basically dishonest piece of play-acting—but then most public functions are.[25]

Scientists are members of a society that has granted them considerable freedom—and money—to pursue knowledge. It is a mutually beneficial arrangement, but it places a serious responsibility on scientists to explain their work, openly and realistically, so society can decide whether to give its informed consent to that research. Unfortunately, a few scientists have used their platforms to offer wild speculations, to raise false hopes or alarms. In the end, everyone loses. Joshua Lederberg has written: "Futuristic pretensions about genetic engineering are a mockery to a mother who has delivered a trisomic [an extra chromosome per cell] child. . . . This disparity between present-day reality and eventual potentiality may arouse deep-seated resentments against that rosier future and even against contemporary scientists who are not quite able to bring it off—in time."[26]

The question of where we should invest our research funds skips past a more basic question: How much of our total resources should we invest in scientific research and development? In the twentieth century we have increasingly looked to science and technology to solve our problems. They have helped us gain great control over our environment. Indeed, we have most of the labor-saving gadgets we need, and many more that we do not. Our homes are ample; there is enough food; clothing is no problem; entertainment comes at the twist of a dial; travel is easy. And tomorrow promises us artificial wombs (another labor-saving device), control over our childrens' sex and genetic features, gadgets to replace all our body parts (except the one that counts most—the brain), and so much more. And in our bargain with technology, we have come to believe that what can be done will be done. But now we must step back and ask a more important, and difficult, question: *Should* it be done?

In the developed countries, we are approaching the limits of technology to bring us happiness. To be sure, we will look to science to develop new energy sources so we can preserve the standard of living we already have. But in a finite world we cannot continue indefinitely to live beyond our means. And we will expect medicine to bring us further freedom from disease. But that will come at a cost—at least for a time—of an increasing population, a rising proportion of old people, and a deterioration of our gene pool.

We face many problems that have no technical solutions, problems we can solve only by changing our values. An artificial womb is not the solution for a couple who want to have "their" child but do not want to postpone their skiing vacation. Amniocentesis plus selective abortion is not the answer for a couple who want a baby girl. ESB will not make life more satisfying. Making people with blond hair, brown eyes, and nice complexions is not the way to make better human beings. Perhaps we should be grateful that our resources are limited, because this forces us to decide, as best we can, what is most worth doing. Indeed, our most serious problems are not physical; they are social and spiritual.

Reverence for Life

Throughout our history we have eaten from the tree of knowledge, and we are becoming as gods. With that knowledge has come more and more power to control our lives, and the lives of others. Indeed, we have traveled far beyond the point of wondering whether to play God; now the question is how to do so wisely.

Are we going too far? Is there some knowledge we should not have? Are there things we should not tamper with? Do we fear what we might learn, that it might upset cherished religious beliefs, or that we may discover we are nothing more than exquisite machines. Leon Kass has written:

We have paid some high prices for the technological conquest of nature, but none perhaps so high as the intellectual and spiritual costs of seeing nature as mere material for our manipulation, exploitation, and transformation. With the powers for biological engineering now gathering, there will be splendid new opportunities for a similar degradation of our view of man. Indeed, we are already witnessing the erosion of our idea of man as something splendid or divine, as a creature with freedom and dignity. And clearly, if we come to see ourselves as meat, then meat we shall become.[27]

But we are far more than meat, and our quest for knowledge cannot obscure the fact that we also are emotional, ethical, social, and spiritual beings. Indeed, there have been deep conflicts in the past between our religious beliefs and our scientific knowledge. At times religion has taken a God-of-the-gaps approach, attributing everything we did not understand to the mysterious workings of God. Consequently, the Church has often resisted new scientific insights, preferring to keep the mysteries of life as the exclusive domain of God. Yet religious beliefs that depend on removable ignorance should not stand in the way. Daniele Petrucci said: "I am a scientist dedicated to uncovering those mysteries of nature which God is prepared to reveal to us."[28]

At the same time, religion has some valid complaints about science. There is a certain arrogance about science, an unspoken implication that the scientific method is the only valid way to seek truth, that it is the only valid way to interpret reality. It is not. Both science and religion are part of our continuous search for greater insight. Science is not the only way, perhaps not even the best, to explore our esthetic, nonrational, and emotional qualities.

Nor can science alone satisfy our search for meaning and purpose in life. Those questions must be pursued by each person, and the answers will not be the same for each of us. No matter how much we learn about its mechanics, life will still hold for us the experiences of mystery and self-discovery. According to Albert Einstein:

> The most beautiful thing we can experience is the mysterious. It is the source of all true art and science. . . . This insight into the mystery of life, coupled though it be with fear, has also given rise to religion. To know that what is impenetrable to us really exists, manifesting itself as the highest wisdom and the most radiant beauty which our dull faculties can comprehend only in their most primitive forms—this knowledge, this feeling is at the center of true religiousness.[29]

The quest for scientific knowledge will go on. But our sense of awe and reverence for life compels us to impose the last, and most important, restriction: respect for the welfare and dignity of each person. People are ends, not means. In 1977 the National Conference of Catholic Bishops adopted the following statement: "A good end or good purpose does not justify any means. There might well be a worthy scientific goal which ought not to be pursued if it unjustifiably violates another human good. In other words, ethical constraints might slow down, or even preclude, some scientific advances."[30]

Knowledge is bought at a price, and sometimes it may be too expensive. For example, it would not be worth the knowledge we would gain to do brain surgery or genetic engineering on healthy people, to abort human fetuses for the sake of an experiment, or to try growing babies entirely in the laboratory before knowing how to do it safely with monkeys.

Our glittering array of tools to manipulate life holds a breathtaking potential for good, and for harm. They offer such compelling benefits that many of us will defend their use for a few specific situations. But each small step will make the next small step easier to accept. And when we reach the top of the stairs, what will we find? What we have most to fear is that we will come to regard people as objects to manufacture and manipulate, all in the name of some noble cause. Stanley Hauerwas of the University of Notre Dame has written: "We must beware of the inhumanity men seem willing to inflict on other men in the name of the good of all men."[31]

This does not have to happen. We do not have to sacrifice our reverence for each person on the altar of new knowledge. But we will have to keep our priorities in order. An informed public—not just the scientists—must decide, from the many things that are technically possible, what will bring us the greatest good. Properly used, our new technologies can enhance the quality of our lives and affirm the worth of each person. Indeed, each of us wants to do what is good, and we need the courage and confidence to seek better answers, even though there are risks.

Our greatest need is not knowledge, but wisdom. And knowledge is less important than people. We need to remember that. This may mean, on occasion, that the development of certain scientific knowledge will just have to wait a while. Perhaps this is not the most efficient way of trying to save society from some of its problems. But it is more important, by far, to have a society worth saving.

References

CHAPTER 1

Opening quotes: Pasternak, B. *Doctor Zhivago*, Pantheon Books, New York, p. 338, 1958; Watson, J. D. *The Double Helix*, Atheneum Publishers, New York p. 197, 1968

1. Ponnamperuma, C. *The Origins of Life*, E. P. Dutton & Co., Inc., New York, p. 21, 1972
2. Cohen, S. N., A. Y. C. Chang, H. W. Boyer, and R. B. Helling. "Construction of biologically functional bacterial plasmids *in vitro*," *Proceedings of the National Academy of Sciences*, U. S., 70: 3240-3244 (1973)
3. "Fully functional gene synthesized at MIT," *Chemical & Engineering News*, pp. 27-30 (Sept. 20, 1976)
4. Sanger, F., G. M. Air, B. G. Barrell, N. L. Brown, A. R. Coulson, J. C. Fiddes, C. A. Hutchison III, P. M. Slocombe, and M. Smith. "Nucleotide sequence of bacteriophage ΦX174 DNA," *Nature*, 265: 687-695 (1977)
5. Silcock, B. "Inside the gene machine," *The Sunday Times* (London), Mar. 6, 1977
6. Crick, F. *Of Molecules and Men*, University of Washington Press, Seattle, p. 87, 1966

CHAPTER 2

Opening quotes: Fletcher, J. *The Ethics of Genetic Control: Ending Reproductive Roulette*, Anchor Press/Doubleday & Company, Garden City, 1974, p. xx; Häring, B. *Ethics of Manipulation*, Seabury Press, New York, 1975, p.196

1. Smith, A. *The Human Pedigree*, J. B. Lippincott Company, Philadelphia, p. 40, 1975
2. Francoeur, R. T. *Utopian Motherhood: New Trends in Human Reproduction*, Doubleday & Company, Garden City, N.Y., p. 19, 1970; Karp, L. E. *Genetic Engineering: Threat or Promise?*, Nelson-Hall, Inc., Chicago, p. 132, 1976
3. Gouldner, H. "Children of the laboratory," *Transaction*, pp. 13-19 (Apr., 1967)
4. Beck, W. J., Jr. "Artificial insemination and semen preservation," *Clinical Obstetrics and Gynecology*, 17: 115-125 (1974); Steinberger, E. and K. D. Smith. "Artificial insemination with fresh or frozen semen," *Journal of the American Medical Association*, 223: 778-783 (1973): Dixon, R. E. and V. C. Buttram, Jr. "Artificial insemination using donor semen: a review of 171 cases,"*Fertility and Sterility*, 27: 130-134 (1976); Jackson, M. C. N. and D. W. Richardson. "The use of fresh and frozen semen in human artificial insemination," *Journal of Biosocial Science*, 9: 251-262 (1977)
5. Beck, "Artificial insemination and semen preservation"; Jackson and Rich-

ardson, "The use of fresh and frozen semen in human artificial insemination"; Dixon, R. E., V. C. Buttram, Jr., and C. W. Schum, "Artificial insemination using homologous semen: a review of 158 cases," *Fertility and Sterility*, 27: 647-654 (1976)

6. Iritani, E. "Sperm banks: 'every man freezes differently,' " *Seattle Post-Intelligencer*, p. C2 (Dec. 7, 1978)

7. "Artificial pregnancy pondered," *Yakima* (Wash.) *Herald-Republic*, p. 10 (Apr. 20, 1975)

8. White, K. "Surrogate mother: relief to childless?" *Seattle Times*, p. K2 (July 9, 1978)

9. Rorvik, D. M. and L. B. Shettles. *Choose Your Baby's Sex*, Dodd, Mead & Company, New York, 1977

10. Guerrero, R. "Association of the type and time of insemination within the menstrual cycle with the human sex ratio at birth," *New England Journal of Medicine*, 291: 1056-1059 (1974)

11. Rorvik and Shettles, *Choose Your Baby's Sex*, pp. 71-72, 127-128; Cornett, S. "Cattlemen may soon pick sex of calves," *Dispatch* (St. Paul, Minn.), p. 26 (Aug. 21, 1975)

12. Ericsson, R. J., C. N. Langevin, and M. Nishino. "Isolation of fractions rich in human Y sperm," *Nature*, 246: 421-424 (1973); Shettles, L. B. "Separation of X and Y spermatozoa," *Journal of Urology*, 116: 462-464 (1976)

13. Dmowski, W. P., L. Gaynor, R. Rao, M. Lawrence, and A. Scommegna. "X and Y separation and clinical experience with AIH separated for male sex preselection," *Proceedings of the First International Symposium on Artificial Insemination Homologous and Male Subfertility*, Bordeaux, France (May 5-8, 1978)

14. Rorvik, D. M. "Embryo transplants," *Good Housekeeping*, pp. 7-8, 124-128 (June, 1975)

15. Strickler, R. C., D. W. Keller, and J. C. Warren. "Artificial insemination with fresh donor semen," *New England Journal of Medicine*, 293: 848-853 (1975); Haman, J. O. "Therapeutic donor insemination," *California Medicine*, 90: 130-133 (1959); Goss, D. A. "Current status of artificial insemination with donor semen," *American Journal of Obstetrics and Gynecology*, 122: 246-252 (1975)

17. Iizuka, R., Y. Sawada, N. Nishina, and M. Ohi. "The physical and mental development of children born following artificial insemination," *International Journal of Fertility*, 13: 24-32 (1968)

18. Iritani, E. "Babies by sperm donors: are fathers a vanishing breed?" *Seattle Post-Intelligencer*, p. C1 (Dec. 3, 1978)

19. Jennings, R. T., R. E. Dixon, and J. B. Nettles. "The risks and prevention of *Neisseria gonorrhoeae* transfer in fresh ejaculate insemination," *Fertility and Sterility*, 28: 554-556 (1977)

20. Fiumara, N. J. "Transmission of gonorrhea by artificial insemination," *British Journal of Venereal Diseases*, 48: 308-309 (1972)

21. Strickler, Keller, and Warren, "Artificial insemination with fresh donor semen"

22. Ratcliff, J. D. "Artificial insemination—has it made happy homes?" *Reader's Digest*, pp. 77–80 (June, 1955)

23. Etzioni, A. "Sex control, science, and society," *Science*, 161: 1107–1112 (1968)

24. Westoff, C. F. and R. R. Rindfuss. "Sex preselection in the United States: some implications," *Science*, 184: 633–636 (1974)

25. Revillard, M. "Legal aspects of artificial insemination and embryo transfer in French domestic law and private international law," in G. E. W. Wolstenholm and D. W. Fitzsimons (editors), *Law and Ethics of A. I. D. and Embryo Transfer*, Ciba Foundation Symposium 17 (new series), Associated Scientific Publishers, Amsterdam, pp. 77–90, 1973

26. Williams, G. *The Sanctity of Life and the Criminal Law*, Alfred A. Knopf, Inc., New York, p. 118, 1970

27. Ibid.

28. White, "Surrogate mother: relief to childless?"

29. Peckins, D. M. "Artificial insemination and the law," *Journal of Legal Medicine*, 4: 17–22 (1976)

30. Seligmann, J. "Life without father," *Newsweek*, p. 87 (Sept. 22, 1975)

31. McLaren, A. "Biological aspects of A. I. D.," in G. E. W. Wolstenholm and D. W. Fitzsimons (editors), *Law and Ethics of A. I. D. and Embryo Transfer*, Ciba Foundation Symposium 17 (new series), Associated Scientific Publishers, Amsterdam, pp. 3–9, 1973

32. Sell, K. W. and H. D. Krause. "Illegitimacy—a professional concern," *Journal of the American Medical Association*, 237: 574 (1977)

33. Wolstenholm, G. E. W., and D. W. Fitzsimons (editors). *Law and Ethics of A.I.D. and Embryo Transfer*, Ciba Foundation Symposium 17 (new series), Associated Scientific Publishers, Amsterdam, p. 63, 1973

34. Fletcher, J. *The Ethics of Genetic Control: Ending Reproductive Roulette*, Anchor Press/Doubleday & Company, Inc., Garden City, N.Y., p. 37, 1974

35. Ramsey, P. "Moral and religious implications of genetic control," in J. D. Roslansky (editor), *Genetics and the Future of Man*, North-Holland Publishing Co., Amsterdam, 1966

36. *The New American Bible*, Catholic Press, Chicago, p. 11, 1971

37. Morris, W. (editor). *The American Heritage Dictionary of the English Language*, American Heritage Publishing Company, Inc. and Houghton Mifflin Company, Boston, p. 18, 1973

38. McLaren, "Biological aspects of A.I.D."; Glotz, P. G. "How to determine the cause of male infertility," *Modern Medicine*, 45: 38–42 (1977)

39. Ratcliff, "Artificial insemination—has it made happy homes?"

CHAPTER 3

Opening quotes: Hafez, E. S. E., as cited in D. Wallechinsky, and I. Wallace, *The People's Almanac*, Doubleday & Company, Inc., Garden City, N.Y., p. 19, 1975; Kass, L. "New beginnings in life," in M. P. Hamilton (editor), *The New*

Genetics and the Future of Man, William B. Eerdmans Publishing Co., Grand Rapids, Mich., pp. 15–63 (1972)

1. Sugie, T., T. Soma, S. Fukumitsu, and K. Otsuki. "Studies on the ovum transfer in cattle, with special reference to collection of ova by means of non-surgical techniques," *Bulletin of the National Institute of Animal Husbandry,* 25: 35 (1972); Oguri, N. and Y. Tsutsumi. "Non-surgical egg transfer in mares," *Journal of Reproduction and Fertility,* 41: 313–320 (1974)

2. Karp, L. E. and R. P. Donahue. "Preimplantation ectogenesis—science and speculation concerning *in vitro* fertilization and related procedures," *Western Journal of Medicine,* 124: 282–298 (1976)

3. Brackett, B. G. "Mammalian fertilization *in vitro,*" *Federation Proceedings,* 32: 2065–2068 (1973)

4. Heape, W. "Preliminary note on the transplantation and growth of mammalian ova within a uterine foster-mother," *Proceedings of the Royal Society* (London), 48: 457–458 (1890)

5. Sugie et al., "Studies on the ovum transfer in cattle"; Oguri and Tsutsumi, "Non-surgical egg transfer in mares."

6. "Monkey motherhood: two for Primero," *Science News,* 109: 5 (Jan. 3, 1976)

7. Kirby, D. R. S. "The transplantation of mouse eggs and trophoblast to extrauterine sites," in J. C. Daniel, Jr. (editor), *Methods in Mammalian Embryology,* W. H. Freeman and Company, San Francisco, pp, 146–156, 1971

8. Whittingham, D. G., S. P. Leibo, and P. Mazur. "Survival of mouse embryos frozen to –196° and –269°C," *Science,* 178: 411–414 (1972)

9. Whittingham, D. G. and W. K. Whitten. "Long-term storage and aerial transport of frozen mouse embryos," *Journal of Reproduction and Fertility,* 36: 433–435 (1974)

10. "English cow gives birth to calf grown from 2d cow's embryo," *New York Times,* p. 14 (June 8, 1973)

11. "The great 'test tube' baby furor—so far," *Medical World News,* pp. 15–16 (Aug. 9, 1974)

12. Steptoe, P. C. and R. G. Edwards. "Reimplantation of a human embryo with subsequent tubal pregnancy," *Lancet,* 1: 880–882 (1976)

13. Gwynne, P., with T. Clifton, M. Hager, S. Begley, and R. Gastel. "All about that baby," *Newsweek,* pp. 66–72 (Aug. 7, 1978)

14. "Calcutta woman has test-tube baby," *Seattle Times,* p. A16 (Oct. 6, 1978)

15. Francoeur, R. T. *Utopian Motherhood: New Trends in Human Reproduction,* Doubleday & Company, Inc., Garden City, N.Y., pp. 57–59, 1970

16. New, D. A. T. and J. C. Daniel, Jr. "Cultivation of rat embryos explanted at 7.5 to 8.5 days of gestation," *Nature,* 223: 515–516 (1969); New, D. A. T. "Studies on mammalian fetuses *in vitro* during the period of organogenesis," in C. R. Austin (editor), *The Mammalian Fetus In Vitro,* Chapman and Hall, London, pp. 15–16, 1973

17. Jenkinson, E. J. and I. B. Wilson. "*In vitro* support system for the study of blastocyst differentiation in the mouse," *Nature,* 228: 776–778 (1970); Hsu, Y. C. "Post-blastocyst differentiation *in vitro,*" *Nature,* 231: 100–102 (1971)

18. Goodlin, R. C. "Foetal incubator," *Lancet*, 1: 1356–1357 (1962); Goodlin, R. C. "An improved fetal incubator," *Transactions of the American Society for Artificial Internal Organs*, 9: 348–350 (1963)

19. Zapol, W. M., T. Kolobow, J. F. Pierce, G. G. Vurek, and R. L. Bowman. "Artificial placenta: two days of total extrauterine support of the isolated premature lamb fetus," *Science*, 166: 617–618 (1969); Zapol, W. M. and T. Kolobow. "Isolated extracorporeal placentation of the fetal lamb," in C. R. Austin (editor), *The Mammalian Fetus In Vitro*, Chapman and Hall, London, pp. 147–193, 1973

20. Zapol and Kolobow, "Isolated extracorporeal placentation of the fetal lamb."

21. Westin, B., R. Nyberg, and G. Enhörning. "A technique for perfusion of the previable human foetus," *Acta Paediatrica*, 47: 339–349 (1958)

22. Walker, C. H. M. and B. J. N. Z. Danesh. "Extracorporeal circulation for the study of the pre-term fetus," in C. R. Austin (editor), *The Mammalian Fetus In Vitro*, Chapman and Hall, London, pp. 209–249, 1973

23. Chamberlain, G. "An artificial placenta," *Journal of Obstetrics and Gynecology*, 100: 615–626 (1968)

24. Umezaki, C., K. P. Katayama, and H. W. Jones, Jr. "Pregnancy rates after reconstructive surgery on the fallopian tubes," *Obstetrics and Gynecology*, 43: 418–424 (1974)

25. Winston, R. M. L. and McClure Browne, J. C. "Pregnancy following autograft transplantation of Fallopian tube and ovary in the rabbit," *Lancet*, 2: 494–495 (1974)

26. Cohen, B. M. "Preliminary experience with vascularized Fallopian tube transplants in the human female," *International Journal of Fertility*, 21: 147–152 (1976); Silló-Seidl, G. "First oviduct transplants," *Acta Europaea Fertilitatis*, 7: 285–298 (1976)

27. Edwards, R. G. "Control of human development," in C. R. Austin and R. V. Short (editors), *Reproduction in Mammals*, Vol. 5, Cambridge University Press, London, pp, 87–113, 1972

28. Jacobs, B. B. "Ovarian allograft survival: prolongation after passage *in vitro*," *Transplantation*, 18: 454–457 (1974); Lueker, D. C. and T. R. Sharpton. "Survival of ovarian allografts following maintenance in organ culture," *Transplantation*, 18: 457–458 (1974)

29. G. E. W. Wolstenholme and D. W. Fitzsimons (editors), *Law and Ethics of A. I. D. and Embryo Transfer*, Ciba Foundation Symposium 17 (new series), Associated Scientific Publishers, Amsterdam, pp. 37–38, 1973

30. Gardner, R. L. and R. G. Edwards. "Control of the sex ratio at full term in the rabbit by transferring sexed blastocysts," *Nature*, 218: 346–348 (1968); "Accelerated breeding of 'superior' cattle is said to be possible," *Wall Street Journal*, p. 20 (Jan. 9, 1976)

31. White, J. J., H. G. Andrews, H. Risemberg, D. Mazur, and J. A. Haller. "Prolonged respiratory support in newborn infants with a membrane oxygenator," *Surgery*, 70: 288–296 (1971)

32. Reynolds, E. O. R. "Management of hyaline membrane disease," *British*

Medical Bulletin, 31: 18–24 (1975); Seligmann, J., with M. Gosnell and D. Shapiro. "New science of birth," *Newsweek*, pp. 55–60 (Nov. 15, 1976)

33. "Ariadne," *New Scientist*, 62: 368 (1974)
34. Shearer, L. "Next: twin tube babies?" *Parade*, pp. 24, 27 (Oct. 8, 1978)
35. Oestreicher, A. "In-vitro fertilization success raises hopes within U.S.," *Family Practice News*, pp. 48–49 (Sept. 1, 1978)
36. "Test tube: halting of experiments results in jury awarding woman $50,000," *Yakima (Wash.) Herald-Republic*, p. 16 (Aug. 19, 1978)
37. Shaw, M. W. and C. Damme. "Legal status of the fetus," in A. Milunsky and G. J. Annas (editors), *Genetics and the Law*, Plenum Press, New York, pp. 3–18, 1976
38. "More on abortion," *Newsweek*, p. 15 (July 12, 1976)
39. Morris, W. (editor). *The American Heritage Dictionary of the English Language*, American Heritage Publishing Co., Inc. and Houghton Mifflin Company, Boston, p. 640, 1973
40. Luria, S. E. *Life: The Unfinished Experiment*, Charles Scribner's Sons, New York, p. 135, 1973
41. Fletcher, J. *The Ethics of Genetic Control: Ending Reproductive Roulette*, Anchor Press/Doubleday & Company, Garden City, N.Y., p. 137, 1974
42. Häring, B. *Medical Ethics*, St. Paul Publications, Slough, pp. 82–85, 1972; Altman, L. K. "Fetal brain said to live at 28 weeks." *New York Times*, p. 36 (May 9, 1975)
43. Karp and Donahue, "Preimplantation ectogenesis"; Thompson, R. S., D. M. Smith, and L. Zamboni. "Fertilization of mouse ova in vitro: an electron microscopic study," *Fertility and Sterility*, 25: 222–249 (1974)
44. Toyoda, Y. and M. C. Chang. "Fertilization of rat eggs *in vitro* by epididymal spermatozoa and the development of eggs following transfer," *Journal of Reproduction and Fertility*, 36: 9–22 (1974)
45. "Test tube: 'more chance of normal baby,' " *Yakima (Wash.) Herald-Republic*, p. 8 (July 25, 1978)
46. Kass, L. R. "Babies by means of in vitro fertilization: unethical experiments on the unborn?" *New England Journal of Medicine*, 285: 1174–1179 (1971)
47. Ramsey, P. *Fabricated Man*, Yale University Press, New Haven, Conn., p. 113, 1970
48. Edwards, R. G. "Mammalian eggs in the laboratory," *Scientific American*, 215(2): 72–81 (1966)
49. Miller, W. B. "Reproduction, technology, and the behavioral sciences [letter]," *Science*, 183: 149 (1974)
50. Lubchenco, L. O., M. Delivora-Papadopoulos, L. J. Butterfield, J. H. French, D. Medcalf, I. E. Hix, J. Danick, J. Dobbs, M. Downs, and E. Freelaw. "Long term follow-up studies of pre-term newborn infants. I. Relationship of handicaps to nursery routines," *Journal of Pediatrics*, 80: 501–508 (1972)
51. Kass, L. R. "New beginnings in life," in M. P. Hamilton (editor), *The New Genetics and the Future of Man*, William B. Eerdmans Publishing Co., Grand Rapids, Mich., p. 34, 1972

52. Karp and Donahue, "Preimplantation ectogenesis."
53. "Mechanical mom,"*Reader's Digest*, p. 113 (Feb., 1974)
54. Grossman, E. "The obsolescent mother: a scenario," *Atlantic*, 227: 39-50 (1971)
55. Hardy, J. T. *Science, Technology and the Environment*, W. B. Saunders Company, Philadelphia, p. 140, 1975

CHAPTER 4

Opening quotes: Huxley, A. *Brave New World*, Harper & Row Publishers, Inc., New York (Perrenial Classic edition), p. 3, 1969; Francoeur, R., as cited in "Man into superman: the promise and peril of the new genetics, *Time*, pp. 33-52 (Apr. 19, 1971)
1. Pincus, G. "The breeding of some rabbits produced by recipients of artificially activated ova," *Proceedings of the National Academy of Sciences, U.S.,* 25: 557-559 (1939)
2. Kaufman, M. H., S. C. Barton, and M. A. H. Surani. "Normal postimplantation development of mouse parthenogenetic embryos to the forelimb bud stage," *Nature*, 265: 53-55 (1977)
3. Wallechinsky, D. and I. Wallace. *The People's Almanac*, Doubleday & Company, Inc., Garden City, N.Y., p. 27, 1975
4. Steward, F. C. "From cultured cells to whole plants: the induction and control of their growth and differentiation," *Proceedings of the Royal Society B*, 175: 1-30 (1970)
5. Bulmer, M. G. *The Biology of Twinning in Man*, Clarendon Press, Oxford, Eng., pp. 7-8, 1970
6. Gardner, E. J. *Principles of Genetics*, 4th edition, John Wiley & Sons, Inc., New York, pp. 331-332, 1972
7. Kelly, S. J. "Studies of the potency of the early cleavage blastomeres of the mouse," in M. Balls and A. E. Wild (editors), *The Early Development of Mammals*, Cambridge University Press, Cambridge, pp. 97-106, 1975
8 Gurdon, J. B. "Transplanted nuclei and cell differentiation," *Scientific American*, 219(6): 24-35 (1968); Gurdon, J. B. and R. A. Laskey. "The transplantation of nuclei from single cultured cells into enucleate frogs' eggs," *Journal of Embryology and Experimental Morphology*, 24: 227-248 (1970); Gurdon, J. B. *The Control of Gene Expression in Animal Development*, Clarendon Press, Oxford, Eng., pp. 28-34, 1974
9. Graham, C. F. "The fusion of cells with one and two cell mouse embryos," *Wistar Institute Symposium*, 9: 19-33 (1969)
10. Bromhall, J. D. "Nuclear transplantation in the rabbit egg," *Nature*, 258: 719-722 (1975)
11. Rorvik, D. *In His Image: The Cloning of a Man*, J. B. Lippincott Co., Philadelphia, 1978
12. Markert, C. L. and R. M. Petters. "Homozygous mouse embryos produced by microsurgery," *Journal of Experimental Zoology*, 201: 295-302 (1977)

13. Hoppe, P. C. and K. Illmensee. "Microsurgically produced homozygous-diploid uniparental mice," *Proceeding of the National Academy of Sciences, U.S.*, 74: 5657–5661 (1977)

14. Anderson, J. "Human clones far in the future," *Seattle Post-Intelligencer*, p. B2 (June 16, 1978)

15. Watson, J. D. "Moving toward the clonal man: is this what we want?" *Atlantic*, 227: 50–53 (1971)

16. Graham, C. F. "The production of parthenogenetic mammalian embryos and their use in biological research," *Biological Reviews*, 49: 399–422 (1974)

17. Gorney, R. *The Human Agenda*, Bantam Books, Inc., New York, p. 220, 1973

18. Gwynne, P. with M. Gosnell, S. Begley, A. Collings, and S. H. Gayle. "All about clones," *Newsweek*, pp. 68–69 (Mar. 20, 1978)

19. Solter, D., W. Biczysko, C. F. Graham, M. Pienkowski, and H. Koprowski. "Ultrastructure of early development of mouse parthenogenones." *Journal of Experimental Zoology*, 188: 1–24 (1974)

20. Linder, D., B. K. McCaw, and F. Hecht. "Parthenogenetic origin of benign ovarian teratomas," *New England Journal of Medicine*, 292: 63–66 (1975)

21. Balfour-Lynn, S. "Parthenogenesis in human beings [letter]," *Lancet*, 1: 1071–1072 (1956)

22. Ramsey, P. "Shall we clone a man?" in K. Vaux (editor), *Who Shall Live?*, Fortress Press, Philadelphia, p. 87, 1970

23. Campbell, J. A. *Chemistry: The Unending Frontier*, Goodyear Publishing Company, Santa Monica, Calif., p. 6, 1978

24. Glass, B. "Science: endless horizons or golden age?" *Science*, 171: 23–29 (1971)

25. Rosenfeld, A. *The Second Genesis: The Coming Control of Life*, Prentice-Hall, Inc., Englewood Cliffs, N.J., pp. 185–186, 1969

CHAPTER 5

Opening quotes: *The Holy Bible* (authorized King James version), Random House, Inc., New York, p. 8, 1943; Muller, H. J. "Man's future birthright," a lecture, University of New Hampshire, 1958, p. 18, as cited in P. Ramsey, "Moral and religious implications of genetic control," in J. D. Roslansky (editor), *Genetics and the Future of Man*, North-Holland Publishing Co., Amsterdam, 1966

1. Lederberg, J. "Biological innovation and genetic intervention," in J. A. Behnke (editor); *Challenging Biological Problems: Directions Toward Their Solution*, Oxford University Press, New York, pp. 7–27, 1972

2. McKusick, V. A. *Mendelian Inheritance in Man: Catalogs of Autosomal Recessive, and X-Linked Phenotypes*, 4th edition, Johns Hopkins Press, Baltimore, 1975

3. Gorney, R. *The Human Agenda*, Bantam Books, Inc., New York, p. 231, 1973; Friedmann, T. "Prenatal diagnosis of genetic disease," *Scientific American*, 225(5): 34–42 (1971); Rubin, S. P. "Genetic counselling: who is at risk?" *The Female Patient*, pp. 44–47, (Aug. 1976)

4. Gorney, *The Human Agenda*; Bender, H. "Is there intelligent life on planet

earth?'' in R. A. Paoletti (editor), *Selected Readings: Genetic Engineering and Bioethics*, 2nd edition, MSS Corporation, New York, pp. 191–195, 1974

5. Fletcher, J. *The Ethics of Genetic Control: Ending Reproductive Roulette*, Anchor Press/Doubleday, Garden City, p. 28, 1974; Heller, J. H. "Human chromosome abnormalities as related to physical and mental dysfunction," *Journal of Heredity*, 60: 239–248 (1969)

6. Jones, A. and W. F. Bodmer. *Our Future Inheritance: Choice or Chance?*, Oxford University Press, London, p. 81, 1974

7. "Man into superman: the promise and peril of the new genetics," *Time*, pp. 33–52 (Apr. 19, 1971)

8. Motulsky, A. G. "Brave New World?" *Science*, 185: 653–663 (1974)

9. Ibid.; Edwards, R. G. "Control of human development," in C. R. Austin and R. V. Short (editors), *Reproduction in Mammals. 5. Artificial Control of Reproduction*, Cambridge University Press, London, pp. 87–113, 1972; Brady, R. O., P. G. Pentchev, and A. E. Gal. "Investigations in enzyme replacement therapy in lipid storage diseases," *Federation Proceedings*, 35: 1310–1315 (1975)

10. Ampola, M. G., M. J. Mahoney, E. Nakamura, and K. Tanaka. "Prenatal therapy of a patient with vitamin-B 12-responsive methylnalonic acidemia," *New England Journal of Medicine*, 293: 313–317 (1975); Harmetz, A. "Medical breakthrough: curing a deadly defect *before* the baby is born," *Today's Health*, pp. 14, 17, 60–62, (Dec., 1974)

11. See *Chemical & Engineering News*, p. 11 (July 16, 1973)

12. "Enzyme helps fight lipid storage disease," *Chemical & Engineering News*, p. 7 (Nov. 18, 1974)

13. Weissmann, G. "Is there an alternative to recombinant DNA?" *The Sciences*, 17: 6–9, 30–31 (1977)

14. Patel, H. M. and B. E. Ryman. "Oral administration of insulin by encapsulation within liposomes," *FEBS Letters*, 62: 60–63 (1976)

15. Snyder, P. D., Jr., F. Wold, R. W. Bernlohr, C. Dullum, R. J. Desnick, W. Krivit, and R. M. Condie. "Enzyme therapy: II. Purified human α–galactosidase A: stabilization to heat and protease degradation by complexing with antibody and by chemical modification," *Biochimica et Biophysica Acta*, 350: 432–436 (1974)

16. Gutte, B. and R. B. Merrifield. "The total snythesis of an enzyme with ribonuclease A activity," *Journal of the American Chemical Society*, 91: 501–502 (1969)

17. Friedmann, T. and R. Roblin. "Gene therapy for human genetic disease?" *Science*, 175: 949–955 (1972)

18. Rowley, P. T. "Genetic screening: whose responsibility?" *Journal of the American Medical Association*, 236: 374–375 (1976)

19. Lamberg, L. "Genetic screening: learning what you never wanted to know," *Today's Health*, 54: 28–31, 53–54 (1976)

20. Kan, Y. W., M. S. Golbus, and R. Trecartin. "Prenatal diagnosis of sickle cell anemia," *New England Journal of Medicine*, 294: 1039–1040 (1976)

21. Motulsky, "Brave New World?"; Jones and Bodmer, "Our Future Inheritance: Choice or Chance?, pp. 69–70

22. Reilly, P. "There's another side to genetic screening," *Prism*, pp. 55–57 (Jan., 1976)

23. Motulsky, "Brave New World?"

24. Powledge, T. M. "New trends in genetic legislation," *The Hastings Center Report*, 3: 6–7 (1973)

25. Nadol, J. "Who shall live? Who shall be aborted? Who shall reproduce? Who shall decide?" *Johns Hopkins Magazine*, pp. 12–17 (May, 1973)

26. Lamberg, "Genetic Screening: Learning what you never wanted to know"; Hook, E. B. "Behavioral implications of the human XYY genotype," *Science*, 179:139–150 (1975)

27. Hook, "Behavioral Implications of the Human XYY Genotype"

28. Kan, Golbus, and Trecartin,"Prenatal Diagnosis of Sickle Cell Anemia"

29. U. K. Collaborative Study. "Maternal serum-alpha-fetoprotein measurement in antenatal screening for anencephaly and spina bifida in early pregnancy," *Lancet*, 1: 1323–1332 (1977)

30. MacVicar, J. "Antenatal detection of fetal abnormality—physical methods," *British Medical Bulletin*, 32: 4–8 (1976)

31. Schwarz, R. H. "Amniocentesis," in *Clinical Obstetrics and Gynecology*, Vol. 18, Harper & Row, Publishers, Inc., New York, pp. 1–22, 1975; Cross, H. E. and A. E. Maumenee. "Ocular trauma during amniocentesis," *Archives of Ophthalmology*, 90: 303–304 (1973)

32. The NICHD National Registry for Amniocentesis Study Group. "Midtrimester amniocentesis for prenatal diagnosis: safety and accuracy," *Journal of the American Medical Association*, 236: 1471–1476 (1976)

33. Epstein, C. J. and M. S. Golbus. "Prenatal diagnosis of genetic diseases," *American Scientist*, 65: 703–711 (1972)

34. McBride, G. "Prenatal diagnosis: problems and outlook," *Journal of the American Medical Association*, 222: 132–135 (1972)

35. Fletcher, *"The Ethics of Genetic Control: Ending Reproductive Roulette*, pp. 56–57

36. Heller, "Human chromosome abnormalities as related to physical and mental dysfunction"

37. Jones and Bodmer, *Our Future Inheritance: Choice or Chance?*, pp. 123–124

38. Nadol, "Who Shall Live? Who Shall Be Aborted? Who Shall Reproduce? Who Shall Decide?"; Fletcher, *The Ethics of Genetic Control: Ending Reproductive Roulette*, p. 137

39. Rhine, S. A., J. L. Cain, R. E. Cleary, C. G. Palmer, and J. F. Thompson. "Prenatal sex detection with endocervical smears: successful results using y-body fluorescence," *American Journal of Obstetrics and Gynecology*, 122: 155–160 (1975); Stattman, E. "Forecasts children's sex," *Ellensburg Wash. Daily Record*, p. 2 (Sept. 6, 1975)

40. Rhine et al., "Prenatal sex detection with endocervical smears."

41. Leff, D. N. "Boy or girl: now choice, not chance," *Medical World News,* pp. 45–56 (Dec. 1, 1975)

42. Etzioni, A. "Selecting the sex of one's children [letter]," *Lancet,* 1: 932–933 (1974)

43. Lazarevic, J. L. "Sex identification of unborn—is it ethical?" *Yakima, Wash. Herald-Republic,* p. 22 (Mar. 3, 1976)

44. McCormick, R. A. "To save or let die—the dilemma of modern medicine," *Journal of the American Medical Association,* 229: 172–176 (1974)

45. Ibid.

46. Ibid.

47. Motulsky, "Brave New World?"; Epstein and Golbus, "Prenatal Diagnosis of Genetic Diseases"; Polani, P. E. and P. F. Benson. "Prenatal diagnosis," *Guy's Hospital Reports,* 122: 65–89 (1973)

48. Winchester, A. M. *Human Genetics,* Charles E. Merrill Publishing Company, Columbus, Ohio, p. 164, 1971

49. McCormick, "To save or let die—the dilemma of modern medicine"; "The hardest choice," *Time,* p. 84 (Mar. 25, 1974)

50. Duff, S. and A. G. M. Campbell. "Moral and ethical dilemmas in the special-care nursery," *New England Journal of Medicine,* 289: 890–894 (1973)

51. "The Hardest Choice"

52. Osborn, F. "The protection and improvement of man's genetic inheritance," in S. Mudd (editor), *The Population Crisis and the Use of World Resources,* Dr. W. Junk Publishers, The Hague, pp. 306–313, 1964

53. Jones and Bodmer, *Our Future Inheritance: Choice or Chance?* p. 20

54. Ramsey, P. "Moral and religious implications of genetic control," in J. D. Roslansky (editor), *Genetics and the Future of Man,* North-Holland Publishing Co., Amsterdam, 1966

55. Friedmann, "Prenatal diagnosis of genetic disease"

56. Motulsky, "Brave new world?"

57. Gorney, *The Human Agenda,* p. 235

58. Smith, J. M. "Eugenics and utopia," *Daedalus,* 94: 487–505 (1965)

59. Fletcher, *The Ethics of Genetic Control: Ending Reproductive Roulette,* p. 196

60. Glass, B. "Science: endless horizons or golden age?" *Science,* 171: 23–29 (1971)

CHAPTER 6

Opening quotes: Sinsheimer, R. "The prospect of designed genetic change," *Engineering and Science* (California Institute of Technology), pp. 8–13 (Apr., 1969); Kelly, W. *Pogo: We Have Met the Enemy and He Is Us,* Simon & Schuster, Inc., New York, 1972

1. Sonneborn, T. M. (editor) *The Control of Human Heredity and Evolution,* The Macmillan Company, New York, p. 37, 1965

2. Jones, A. and W. F. Bodmer. *Our Future Inheritance: Choice or Chance?* Oxford University Press, London, p. 109, 1974

3. Muller, H. J. "The gene material as the initiator and the organizing basis of

life," in R. A. Brink (editor), *Heritage from Mendel,* University of Wisconsin Press, Madison, pp. 419–447, 1967; Lewin, B. *Gene Expression. 2. Eucaryotic Chromosomes,* John Wiley & Sons, London, p. 149, 1974

4. Harris, H. and J. F. Watkins. "Hybrid cells derived from mouse and man: artificial heterokaryons of mammalian cells from different species," *Nature,* 205: 640–646 (1965)

5. Siniscalco, M., H. P. Klinger, H. Eagle, H. Koprowski, W. Y. Fujimoto, and J. E. Seegmiller. "Evidence for intragenic complementation in hybrid cells derived from two human diploid strains each carrying an X-linked mutation," *Proceedings of the National Academy of Sciences, U.S.,* 62: 793–799 (1969); Silagi, S., G. Darlington, and S. Bruce. *"Hybridization of two biochemically marked cell lines." Proceedings of the National Academy of Sciences. U.S.,* 62: 1085–1092 (1969)

6. Weiss, M. and H. Green. "Human-mouse hybrid cell lines containing partial complements of human chromosomes and functioning human genes," *Proceedings of the National Academy of Sciences, U.S.,* 58: 1104–1111 (1967)

7. Zepp, H. D., J. H. Conover, K. Hirschhorn, and H. L. Hodes. "Human-mosquito somatic cell hybrids induced by ultraviolet-inactivated Sendai virus," *Nature,* 229: 119–121 (1971)

8. Schwartz, A. G., P. R. Cook, and H. Harris. "Correction of a genetic defect in a mammalian cell," *Nature,* 230: 5–8 (1971)

9. Sawicki, W. and H. Koprowski. "Fusion of rabbit spermatozoa with somatic cells cultivated *in vitro,*" *Experimental Cell Research,* 66: 145–151 (1971); Phillips, S. G., D. M. Phillips, V. G. Dev, D. A. Miller, O. P. Van Diggelen, and O. J. Miller. "Spontaneous cell hybridization of somatic cells present in sperm suspensions," *Experimental Cell Research,* 98: 429–443 (1976)

10. Boyd, Y. L. and H. Harris. "Correction of genetic defects in mammalian cells by the input of small amounts of foreign genetic material," *Journal of Cell Science,* 13: 841–861 (1973); Ruddle, F. H. and R. P. Creagan. "Parasexual approaches to the genetics of man," in H. L. Roman, A. Campbell, and L. M. Sandler (editors), *Annual Review of Genetics,* Vol. 9, Annual Reviews Inc., Palo Alto, Calif., pp. 407–486, 1975

11. Earley, K. "Disinheriting disease," *The Sciences,* 15: 19–23 (1975)

12. McBride, O. W. and H. L. Ozer. "Transfer of genetic information by purified metaphase chromosomes," *Proceedings of the National Academy of Sciences, U.S.,* 70: 1258–1262 (1973)

13. Willecke, K. and F. H. Ruddle. "Transfer of the gene for hypoxanthine guanine phosphoribosyltransferase on isolated human metaphase chromosomes into murine L-cells," *Proceedings of the National Academy of Sciences, U.S.,* 72: 1792–1796 (1975); Burch, J. W. and O. W. McBride. "Human gene expression in rodent cells after uptake of isolated metaphase chromosomes," *Proceedings of the National Academy of Sciences, U.S.,* 72: 1797–1801 (1975)

14. Cohen, S. N., A. Y. C. Chang, H. W. Boyer, and R. B. Helling. "Construction of biologically functional bacterial plasmids *in vitro,*" *Proceedings of the National Academy of Sciences, U. S.,* 70: 3240–3244 (1973)

15. Cohen, S. N. "The manipulation of genes," *Scientific American*, 233(1): 24–33 (1975); Williamson, B. "First mammalian results with genetic recombinants," *Nature*, 260: 189–190 (1976); Lippincott, W. T. "Recombinant DNA: how much risk is too much?" *Journal of Chemical Education*, 54: 265 (1977)

16. Fernandez, S. M., P. F. Lurquin, and C. I. Kado. "Incorporation and maintenance of recombinant-DNA plasmid vehicles pBR313 and pCR1 in plant protoplasts," *FEBS (Federation of European Biochemical Societies) Letters*, 87: 277–282 (1978); Fox, J. L. "Plant genetic engineering work advances," *Chemical & Engineering News*, pp. 17–18 (July 17, 1978); Stanfield, S. and D. R. Helinski. "Small circular DNA in *Drosophila melanogaster*," *Cell*, 9: 335–345 (1976); Nonoyama, M. and A. Tanaka. "Plasmid DNA as a possible state of Epstein-Barr virus genomes in nonproductive cells," *Cold Spring Harbor Symposia on Quantitative Biology*, 39: 807–810 (1975)

17. Shapiro, J., L. MacHattie, L. Eron, G. Ihler, K. Ippen, and J. Beckwith. "Isolation of pure lac operon DNA," *Nature*, 224: 768–774 (1969)

18. Agarwal, K. L., H. Buchi, M. H. Caruthers, N. Gupta, H. G. Khorana, K. Kleppe, A. Kumar, E. Ohtsuka, U. L. Rajbhandary, J. H. Van de Sande, V. Sgaramella, H. Wever, and T. Yamada. "Total synthesis of the gene for an alanine transfer ribonucleic acid from yeast," *Nature*, 227: 27–34 (1970)

19. "Fully functional gene synthesized at MIT," *Chemical & Engineering News*, pp. 27–30 (Sept. 20, 1976)

20. Williamson, "First mammalian results with genetic recombinants"; Maniatis, T., S. G. Kee, A. Efstratiadis, and F. C. Kafatos. "Amplification and characterization of a β-globin gene synthesized *in vitro*," *Cell*, 8: 163–182 (1976); "Rabbit gene works in monkey cells," *Chemical & Engineering News*, p. 8 (Oct. 30, 1978); Ullrich, A., J. Shine, J. Chirgwin, R. Pictet, E. Tischer, W. J. Rutter, and H. W. Goodman. "Rat insulin genes: construction of plasmids containing the coding sequences," *Science*, 196: 1313–1319 (1977)

21. Thomas, L. *The Lives of a Cell*, The Viking Press, New York, p. 4, 1974

22. Struhl, K., J. R. Cameron, and R. W. Davis. "Functional genetic expression of eucaryotic DNA in *Escherichia coli*," *Proceedings of the National Academy of Sciences, U.S.*, 73: 1471–1475 (1976)

23. Doy, C. H., P. M. Gresshof, and B. G. Rolfe. "Biological and molecular evidence for the transgenosis of genes from bacteria to plant cells," *Proceedings of the National Academy of Sciences, U.S.*, 70: 723–726 (1973); "Rabbit gene works in monkey cells"; Merril, C. R., M. R. Geier, and J. C. Petricciani. "Bacterial virus gene expression in human cells," *Nature*, 233: 398–400 (1971)

24. Vogel, F. and R. Rathenberg. "Spontaneous mutation in man," in H. Harris and K. Kirschhorn (editors), *Advances in Human Genetics*, Vol. 5, Plenum Press, New York, pp. 223–318, 1975

25. Burnet, M. *Genes, Dreams and Realities*, Medical and Technical Publishing Co. Ltd., Aylesbury, Eng., pp. 9–13, 1971

26. Mintz, B. "Allophenic mice of multi-embryo origin," in J. C. Daniel (editor), *Methods in Mammalian Embryology*, W. H. Freeman and Company, San Francisco, pp. 186–214, 1971

27. Mintz, B. "Gene control of mammalian differentiation," in H. L. Roman, A. Campbell, and L. M. Sandler (editors), *Annual Review of Genetics*, Vol. 8, Annual Reviews Inc., Palo Alto, Calif., pp. 411–470, 1974

28. Chaleff, R. S. and P. S. Carlson. "Somatic cell genetics of higher plants," in H. L. Roman, A. Campbell, and L. M. Sandler (editors), *Annual Review of Genetics*, Vol. 8, Annual Reviews Inc., Palo Alto, Calif., pp. 267–278, 1974

29. McLaren, A. *Mammalian Chimaeras*, Cambridge University Press, Cambridge, p. 7, 1976

30. Illmensee, K., P. C. Hoppe, and C. M. Croce. "Chimeric mice derived from human-mouse hybrid cells," *Proceedings of the National Academy of Sciences, U.S.*, 75: 1914–1918 (1978)

31. Mintz, B. and K. Illmensee. "Normal genetically mosaic mice produced from malignant teratocarcinoma cells," *Proceedings of the National Academy of Sciences, U.S.*, 72: 3585–3589 (1975)

32. "Mouse chimeras help study muscle disease," *Chemical & Engineering News*, p. 20 (Nov. 13, 1978)

33. "Sickle-cell test based on restriction enzymes," *Chemical & Engineering News*, p. 20 (Nov. 13, 1978)

34. Carlson, P. S., H. H. Smith, and R. D. Dearing. "Parasexual interspecific plant hybridization," *Proceedings of the National Academy of Sciences, U.S.*, 69: 2292–2294 (1972)

35. Gore, R. "The awesome worlds within a cell," *National Geographic*, pp. 355–395 (Sept., 1976)

36. Dixon, R., F. C. Cannon, and A. Kondorosi. "Construction of a P plasmid carrying nitrogen fixation genes from *Klebsiella pneumoniae*," *Nature*, 260: 268–271 (1976); Cannon, F. C. and J. R. Postgate. "Expression of *Klebsiella* nitrogen fixation genes (nif) in *Azotobacter*," *Nature*, 260: 271–272 (1976)

37. Itakura, K., T. Hirose, R. Crea, A. D. Riggs, H. L. Heyneker, F. Bolivar, and H. W. Boyer. "Expression in *Escherichia coli* of a chemically synthesized gene for the hormone somatostatin," *Science*, 198: 1056–1063 (1977)

38. Gunby, P. "Bacteria directed to produce insulin in test application of genetic code," *Journal of the American Medical Association*, 240: 1697–1698 (1978)

39. Sinsheimer, R. "An evolutionary perspective for genetic engineering," *New Scientist*, 73: 150–152 (1977)

40. National Institutes of Health. "Recombinant DNA research guidelines: draft environmental impact statement," *Federal Register*, 41(176): 38426–38483 (1976)

41. Davis, B. D. "The recombinant DNA scenarios: Andromeda strain, Chimera, and Golem," *American Scientist*, 65(5): 547–555 (1977)

42. O'Sullivan, D. A. "Genetic engineering to resume—cautiously," *Chemical & Engineering News*, p. 19 (Mar. 10, 1975)

43. Hayflick, L. "The limited *in vitro* lifetime of human diploid cell strains," *Experimental Cell Research*, 37: 614–636 (1965)

44. Friedmann, T. and R. Roblin. "Gene therapy for human genetic disease?" *Science*, 175: 949–955 (1972)

45. Ibid.; Earley, "Disinheriting disease."

46. Fox, M. S. and J. W. Littlefield. "Reservations concerning gene therapy," *Science,* 173: 195 (1971)
47. Jones and Bodmer, *Our Future Inheritance: Choice or Chance?,* p. 114
48. Lederberg, J. "Biological innovation and genetic intervention," in J. A. Behnke (editor), *Challenging Biological Problems: Directions Toward Their Solution,* Oxford University Press, New York, pp. 7–27, 1972
49. Snow, C. P. "Human care," *Journal of the American Medical Association,* 225: 617–621 (1973)
50. Teilhard de Chardin, P. *The Phenomenon of Man,* Harper & Row Publishers, Inc., New York, p. 249, 1959
51. Wade, N. *The Ultimate Experiment: Man-Made Evolution,* Walker and Company, New York, p. 4, 1977
52. Kass, L. R. "The new biology revives vital concerns," *Seattle Post-Intelligencer,* p. D12 (Feb. 3, 1974)
53. Sinsheimer, R. "The brain of Pooh," *Engineering and Science* (California Institute of Technology), Jan. 1970

CHAPTER 7

Opening quotes: Harrington, A. *The Immortalist,* Random House, Inc., New York, 1969, p. 3; Seneca, L. A., as cited in G. Dunea, "Death with dignity," *British Medical Journal,* 1: 824–825 (1976)
1. Rosenfeld, A. *Prolongevity,* Alfred A. Knopf, New York, p. 1, 1976
2. Young, J. Z. *An Introduction to the Study of Man,* Oxford University Press, London, p. 297, 1974
3. Kent, S. "How do we age?", *Geriatrics,* 31: 128–134 (1976)
4. Strehler, B. L. "The understanding and control of the aging process," in J. A. Behnke (editor), *Challenging Biological Problems: Directions Toward Their Solution,* Oxford University Press, New York, pp. 133–147, 1972
5. Ibid.; Brandes, D. "Lysosomes and aging pigment," in P. L. Krohn (editor), *Topics in the Biology of Aging,* John Wiley & Sons, Inc., New York, pp. 149–157, 1966
6. Curtis, H. J. "The role of somatic mutations in aging," in P. L. Krohn (editor), *Topics in the Biology of Aging,* John Wiley & Sons, Inc., New York, pp. 63–74, 1966
7. Orgel, L. E. "The maintenance of the accuracy of protein synthesis and its relevance to aging," *Proceedings of the National Academy of Sciences, U.S., 49:* 517–521 (1963)
8. Sanders, H. J. "Human aging: the enigma persists," *Chemical & Engineering News,* pp. 13–16 (July 24, 1972)
9. Strehler, "The Understanding and Control of the Aging Process."
10. Walford, R. L. "Generalizing biologic hypotheses and aging: an immunological approach," in P. L. Krohn (editor), *Topics in the Biology of Aging,* John Wiley & Sons, Inc., New York, pp. 163–169, 1966
11. Sanders, "Human Aging: the enigma persists"

12. "Scientists weigh facts, theories on aging," *Chemical & Engineering News*, pp. 14–15 (Mar. 18, 1974); Pryor, W. A. "Free radical pathology," *Chemical & Engineering News*, pp. 34–51 (June 7, 1971)

13. Sanders, "Human aging: the enigma persists"; Rosenfeld, A., *Prolongevity*, pp. 106–116

14. McCay, C. M., L. A. Maynard, G. Sperling, and L. L. Barnes. "Retarded growth, life span, ultimate body size and age changes in the albino rat after feeding diets restricted in calories," *Journal of Nutrition*, 18: 1–13 (1939)

15. Sanders, "Human Aging: the enigma persists"

16. Krohn, P. L. "Transplantation and aging," in P. L. Krohn (editor), *Topics in the Biology of Aging*, John Wiley & Sons, Inc., New York, pp. 125–139, 1966

17. Hayflick, L. "The limited *in vitro* lifetime of human diploid cell strains," *Experimental Cell Research*, 37: 614–636 (1965)

18. Orgel, L. E. "Ageing of clones of mammalian cells," *Nature*, 243: 441–445 (1973)

19. Comfort, A. "Basic research in gerontology," *Gerontologia*, 16: 48–64 (1970)

20. Wallechinsky, D. and I. Wallace. *The People's Almanac*, Doubleday and Company, Inc., Garden City, N.Y., p. 19, 1975

21. Sanders, "Human aging: The enigma persists"

22. Strehler, "The understanding and control of the aging process"

23. Taylor, G. R. *The Biological Time Bomb*, New American Library, New York, p. 95, 1968

24. "Can aging be cured?" *Newsweek*, pp. 56–66 (Apr. 16, 1973)

25. Smith, J. M. "Eugenics and utopia," *Daedalus*, 94: 487–505 (1965)

26. "Kidneys from cadavers," *British Medical Journal*, 1: 188 (1977)

27. Calne, R. Y. "Mechanisms in the acceptance of organ grafts," *British Medical Bulletin*, 32: 107–112 (1976); Fishlock, D. *Man Modified*, Jonathan Cape, London, p. 162, 1969; Clark, M. with M. Hager and J. Huck. "New look at transplants," *Newsweek*, pp. 63–64 (July 31, 1978)

28. Clark with Hager and Huck, "New look at transplants"; Calne, R. Y. "The present status of liver transplantation," *Transplantation Proceedings*, 9: 209–216 (1977)

29. Clark with Hager and Huck, "New look at transplants"; Sandiford, D. M., N. E. Shumway, E. B. Stinson, and B. Reitz. "Cardiac transplantation: eight years' experience," *Heart & Lung*, 566–570 (1976)

30. Fishlock, D., *Man Modified*, p. 168

31. "The sale of human body parts," *Michigan Law Review*, 72: 1182–1264 (1974)

32. Linask, J., J. Votta, and M. Willis. "Perfusion preservation of hearts for 6 to 9 days at room temperature," *Science*, 199: 299–301 (1978)

33. Rapatz, G. "Resumption of activity in frog hearts after freezing to low temperatures," *Biodynamica*, 11: 1–12 (1970)

34. Offerijns, F. G. J. and H. W. Krijnen. "The preservation of the rat heart in the frozen state," *Cryobiology*, 9: 289–295 (1972)

35. Mazur, P., J. A. Kemp, and R. H. Miller. "Survival of whole fetal rat pancreases frozen to –78 and –196°C," *Cryobioloby*, 13: 647 (1976)

36. "Souls on ice," *Newsweek*, p. 11 (Aug. 16, 1976)

37. Wiley, J. P., Jr. and J. K. Sherman. "Immortality and the freezing of human bodies," *Natural History*, 80 (10): 12–22 (1971)

38. Reetsma, K. "Heterotransplantation," in F. T. Rapaport and J. Dausset (editors), *Human Transplantation*, Grune & Stratton, New York, pp. 357–366, 1968

39. Seligmann, J. with P. Young-Husband. "The baboon heart," *Newsweek*, p. 60 (July 4, 1977); "Man receives chimpanzee's heart in transplant operation," *Ellensburg* (Wash.) *Daily Record*, p. 8 (Oct. 14, 1977)

40. Saunders, S. J., J. Terblanche, S. C. W. Bosman, G. G. Harrison, R. Walls, R. Hickman, J. Biebuyck, D. Dent, S. Pearce, and C. N. Barnard. "Acute hepatic coma treated by cross-circulation with a baboon and by repeated exchange transfusions," *Lancet*, 2: 585–588 (1968)

41. Goss, R. J. *Principles of Regeneration*, Academic Press, New York, pp. 217–219, 1969

42. Ibid., p. 276

43. Wallechinsky and Wallace, *The People's Almanac*, pp. 18–19

44. Povlson, C. O., N. E. Skakkebaek, J. Rygaard, and G. Jensen. "Heterotransplantation of human foetal organs to the mouse mutant *nude*," *Nature*, 248: 247–249 (1974)

45. Sanders, H. J. "Artificial organs, Part 1," *Chemical & Engineering News*, pp. 32–49 (Apr. 5, 1971)

46. "The modern men of parts," *Time*, pp. 73–75 (Mar. 18, 1974)

47. Clark, M. with D. Shapiro and M. Hager. "Do-it-yourself kidney care," *Newsweek*, p. 72 (Mar. 20, 1978); "The body may be best," *Time*, p. 82 (Dec. 18, 1978)

48. "Progress with an artificial liver," *Lancet*, 2: 992–994 (1974)

49. "Device may lead to artificial liver," *Chemical & Engineering News*, p. 49 (Sept. 25, 1978)

50. "Artificial device for diabetes could eventually help humans," *Sioux Falls* (S. Dak.) *Argus-Leader*, p. 4 (Aug. 11, 1976); Albisser, A. M., B. S. Leibel, T. G. Ewart, Z. Davidovac, C. K. Botz, W. Zingg, H. Schipper, and R. Gander. "Clinical control of diabetes by the artificial pancreas," *Diabetes*, 23: 397–404 (1974)

51. "Researchers pull plug on ailing calf," *Ellensburg* (Wash.) *Daily Record*, p. 10 (Nov. 8, 1978)

52. Rossiter, A., Jr. "Blood pump helps save life," *Ellensburg* (Wash.) *Daily Record*, p. 4 (Dec. 1, 1977)

53. Comfort, A. "Longevity of man and his tissues," in G. Wolstenholme (editor), *Man and His Future*, J. & A. Churchill Ltd., London, pp. 217–229, 1963

54. "The sale of human body parts."

55. Hamburger, J. and J. Crosnier. "Moral and ethical problems in transplantation," in F. T. Rapaport and J. Dausset (editors), *Human Transplantation*, Grune & Stratton, New York, pp. 37–44, 1968; Hume, D. M. "Kid-

23. Ibid., pp. 91–92
24. Pines, M., *The Brain Changers*, p. 85
25. Calder, *The Mind of Man*, p. 111
26. McConnell, J. V. "Memory transfer through cannibalism in planarians," *Journal of Neuropsychiatry*, 3: S42–S48 (1962)
27. Hyden, H. and E. Egyhazi. "Changes in RNA content and base composition in cortical neurons of rats in a learning experiment involving transfer of handedness," *Proceedings of the National Academy of Sciences, U.S.*, 52: 1030–1035 (1964)
28. "Man into superman: the promise and peril of the new genetics," *Time*, pp. 33–52 (Apr. 19, 1971)
29. "Chemical theory of memory gains support," *Chemical & Engineering News*, p. 20 (Nov. 26, 1973)
30. Ibid.; "Small peptide induces fear of darkness in rats," *Chemical & Engineering News*, pp. 27–28 (Jan. 11, 1971)
31. Pribram, K. H. "The neurophysiology of remembering," *Scientific American*, 220: 73–86 (1969)
32. Villet, "Opiates of the mind"
33. Leach, *The Biocrats*, pp. 189–190
34. Zinser, B. "Drug increases learning ability," *Dispatch* (Saint Paul, Minn.), Aug. 24, 1976
35. Davis, K. L., R. C. Mohs, J. R. Tinklenberg, A. Pfefferbaum, L. E. Hollister, and B. S. Kopell. "Physostigmine: improvement of long-term memory processes in normal humans," *Science*, 201: 272–274 (1978); Sitaram, N., H. Weingartner, and J. C. Gillin. "Human serial learning: enhancement with arecholine and choline and impairment with scopalamine," *Science*, 201: 274–276 (1978)
36. Sanders, H. J. "Human aging: the enigma persists," *Chemical & Engineering News*, pp. 13–16 (July 24, 1972)
37. Pines, *The Brain Changers*, p. 223
38. Zamenhof, S., E. Van Marthens, and L. Grauel. "Prenatal cerebral development: effect of restricted diet, reversal by growth hormone," *Science*, 174: 954–955 (1971); Clark, G. M., S. Zamenhof, E. Van Marthens, L. Grauel, and L. Kruger. "The effect of prenatal malnutrition on dimensions of the cerebral cortex," *Brain Research*, 54: 397–402 (1973); Zamenhof, S., E. Van Marthens, and L. Grauel. "DNA (cell number) and protein in neonatal rat brain: alteration by timing of maternal dietary protein restriction." *Journal of Nutrition*, 101: 1265–1270 (1971); Zamenhof, S., S. M. Hall, L. Grauel, E. Van Marthens, and M. J. Donahue. "Deprivation of amino acids and prenatal brain development in rats," *Journal of Nutrition*, 104: 1002–1007 (1974)
39. Simonson, M. and B. F. Chow. "Maze studies on progeny of underfed mother rats," *Journal of Nutrition*, 100: 685–690 (1970); Hsueh, A. M., M. Simonson, B. F. Chow, and H. M. Hanson "The importance of the period of dietary restriction of the dam on behavior and growth in the rat," *Journal of Nutrition*, 104: 37–46 (1974)
40. Calder, *The Mind of Man*, p. 240

41. Howard, T. and J. Rifkin. *Who Should Play God?*, Dell Publishing Co., Inc., New York, p. 157, 1977

42. Zamenhof, Marthens, and Gravel, "Prenatal cerebral development"; Chow, B. F. and C. J. Lee. "Effect of dietary restriction of pregnant rats on body weight gain of the offspring," *Journal of Nutrition*, 82: 10–18 (1964)

43. Taylor, G. R. *The Biological Time Bomb*, Panther Books Ltd., London, pp. 155–156, 1969; Rorvik, D. M. and O. S. Heyns. *Decompression Babies*, Dodd, Mead & Company, New York, 1973

44. Hubel, D. H. and T. N. Wiesel. "The period of susceptibility to the physiological effects of unilateral eye closure in kittens," *Journal of Physiology*, 206: 419–436 (1970)

45. Blakemore, C. and G. F. Cooper. "Development of the brain depends on the visual environment," *Nature*, 228: 477–478 (1970)

46. Bennett, E. L., M. C. Diamond, D. Krech, and M. R. Rosenzweig. "Chemical and anatomical plasticity of brain," *Science*, 146: 610–619 (1964)

47. Fiala, B. A., J. N. Joyce, and W. T. Greenough. "Environmental complexity modulates growth of granule cell dendrites in developing but not adult hippocampus of rats," *Experimental Neurology*, 59: 372–383 (1978)

48. Panati, C. with P. Gwynne. "The futurologists," *Newsweek*, p. 52 (Mar. 8, 1976)

49. Thomas, L. *The Lives of a Cell*, The Viking Press, New York, pp. 132–133, 1974

50. "Minority speakers criticize researchers," *Journal of the American Medical Association*, 235: 462 (1976)

51. "Exploring the frontiers of the mind"

52. Pines, *The Brain Changers*, p. 227

53. Stevens, *Explorers of the Brain*, pp. 285–287

54. Pines, *The Brain Changers*, p. 229

55. "Minority speakers criticize researchers"

56. Pines, *The Brain Changers*, pp. 40–41

57. Lederberg, J. "Genetic engineering, or the amelioration of genetic defect," *Pharos*, 34: 9–12 (1971)

58. Valenstein, E. S. *Brain Control*, John Wiley & Sons, New York, pp. 233–236, 1973

59. Mark, V. and F. Ervin. *Violence and the Brain*, Harper & Row, Publishers, New York, 1970

60. Fernald, L. D. and P. S. Fernald. *Introduction to Psychology*, 4th edition, Houghton Mifflin Company, Boston, p. 11, 1978

61. Hardy, J. T. *Science, Technology and the Environment*, W. B. Saunders Company, Philadelphia, p. 126, 1975

62. Fishlock, D. *Man Modified*, Jonathan Cape Ltd., London, p. 82, 1969

63. Smith, M. "When psychology grows up," *New Scientist*, 64: 90–93 (1974)

64. "Man into superman: The promise and the peril of the new genetics"

65. Calder, *The Mind of Man*, p. 215

CHAPTER 9

Opening quotes: Heilbroner, R. L., as cited in Hardy, J. T. *Science, Technology and the Environment,* W. B. Saunders Co., Philadelphia, p. 318 1975; Mao Tse-tung, as cited in Myers, N. " 'Of all things people are the most precious,' " *New Scientist,* 65: 56–59 (1975)

1. Lippincott, W. T. "Recombinant DNA: how much risk is too much?" *Journal of Chemical Education,* 54: 265 (1977)
2. Elwell, L. P. and S. Falkow, "Genetic loose change," *The Sciences,* 17: 8–11 (1977)
3. Abelson, J. "Recombinant DNA: examples of present-day research," *Science,* 196: 159–160 (1977)
4. Gwynne, P. with S. G. Michaud and W. J. Cook. "Politics and genes," *Newsweek,* pp. 50–52 (Jan. 12, 1976)
5. "Tinkering with life," *Time,* pp. 46–51 (Apr. 18, 1977)
6. Culliton, B. J. "Kennedy: pushing for more public input in research," *Science,* 188: 1187–1189 (1975)
7. "Patents: a right to life," *Newsweek,* p. 71 (Mar. 13, 1978)
8. Culliton, B. J. "Recombinant DNA: Cambridge City Council votes moratorium," *Science,* 193: 300–301 (1976)
9. Garrigan, O. *Man's Intervention in Nature,* Hawthorn Books, Inc., New York, p. 7, 1967
10. "NAS's Handler urges scientific freedom," *Chemical & Engineering News,* p. 7 (Oct. 18, 1976)
11. Zuckerman, S. (former chief science advisor in Great Britain), as quoted in "Science, politics relationship discussed," *Chemical & Engineering News,* pp. 18–19 (Mar. 7, 1977)
12. Davis, B. D. "Prospects for genetic intervention in man," *Science,* 170: 1279–1283 (1970)
13. Lederberg, J. "Gene splicing: will fear rob us of its benefits?" *Prism,* pp. 33–37 (Nov., 1975)
14. Davis, "Prospects for genetic intervention in man"
15. Davis, B. D. "Potential benefits are large, protective methods make risks small," *Chemical & Engineering News,* pp. 27–31 (May 30, 1977)
16. Potter, V. R. *Bioethics: Bridge to the Future,* Prentice-Hall, Inc., Englewood Cliffs, N. J., pp. 70, 72, 1971
17. Stern, C. "Genes and people," in A. S. Baer (editor), *Heredity and Society: Readings in Social Genetics,* The Macmillan Company, New York, pp. 311–318, 1973
18. Davis, B. D. (microbiologist at Harvard University), as quoted in Gwynne with Michaud and Cook, "Politics and genes"
19. Baumiller, R. C. (of the Kennedy Institute's Center for Bioethics), as quoted in Gwynne with Michaud and Cook, "Politics and genes"
20. Hardin, G. "The tragedy of the commons," *Science,* 162: 1243–1248 (1968)

21. "Recombinant DNA revisited," *Medical Tribune*, p. 11 (Sept. 7, 1977)
22. Krimsky, S. "Public must regulate recombinant research," *Chemical & Engineering News*, pp. 36–39 (May 30, 1977)
23. Fishlock, D. *The Business of Science*, John Wiley & Sons, New York, p. 23, 1975
24. "User survey okays NSF peer review system," *Chemical & Engineering News*, pp. 16, 21 (May 30, 1977)
25. Burnet, M. *Genes, Dreams and Realities*, Basic Books, Inc., New York, p. 145, 1971
26. Lederberg, J. "Biological innovation and genetic intervention," in J. A. Behnke (editor), *Challenging Biological Problems: Directions Toward Their Solutions*, Oxford University Press, New York, pp. 7–27, 1972
27. Kass, L. R. "Making babies—the new biology and the 'old' morality," *Public Interest*, pp. 18–56 (Winter, 1972)
28. Francoeur, R. T. *Utopian Motherhood: New Trends in Human Reproduction*, Doubleday & Company, Garden City, N.Y., p. 81, 1970
29. Liechty, R. D. "The world's one crime," *Journal of the American Medical Association*, 235: 76 (1976)
30. Wade, N. *The Ultimate Experiment: Man-Made Evolution*, Walker and Company, New York, pp. 143–144, 1977
31. Hauerwas, S. "The meaning of being human," in R. A. Paoletti (editor), *Selected Readings: Genetic Engineering and Bioethics*, 2nd edition, MSS Information Corporation, New York, pp. 195–199, 1974

Suggested Readings

BOOKS

Baer, A. S. (editor). *Heredity and Society: Readings in Social Genetics*, 2nd edition, The Macmillan Company, New York, 1977

Behnke, J. A. (editor). *Challenging Biological Problems: Directions Toward Their Solution*, Oxford University Press, New York, 1972

Burnet, M. *Genes, Dreams and Realities*, Basic Books, Inc., New York, 1971

Calder, N. *The Mind of Man: An Investigation into Current Research on the Brain and Human Nature*, Viking Press, New York, 1971

Crick, F. *Of Molecules and Men*, University of Washington Press, Seattle, Wash., 1966

Fishlock, D. *Man Modified*, Jonathan Cape Ltd., London, 1969

Fletcher, J. *The Ethics of Genetic Control: Ending Reproductive Roulette*, Anchor Press/Doubleday, Garden City, N. Y., 1974

Francoeur, R. T. *Utopian Motherhood: New Trends in Human Reproduction*, Doubleday & Company, Garden City, N.Y., 1970

Goodfield, J. *Playing God*, Random House, Inc., New York, 1977

Gorney, R. *The Human Agenda*, Simon & Schuster, Inc., New York, 1972

Gorovitz, S. (editor). *Moral Problems in Medicine*, Prentice-Hall, Inc., Englewood Cliffs, N.J., 1976

Hamilton, M. P. (editor). *The New Genetics and the Future of Man*, William B. Eerdmans Publishing Co., Grand Rapids, Mich., 1972

Häring, B. *Medical Ethics*, St. Paul Publications, Slough, 1972

Howard, T. and J. Rifkin *Who Should Play God?*, Dell Publishing Co., Inc., New York, 1977

Humber, J. H. and R. F. Almeder (editors). *Biomedical Ethics and the Law*, Plenum Press, New York, 1976

Hutton, R. *Bio-Revolution: DNA and the Ethics of Man-Made Life*. Mentor/New American Library, Bergenfield, N.J., 1978

Huxley, A. *Brave New World*, Harper & Row, Publishers, Inc., New York, 1969

Jones, A. and W. F. Bodmer. *Our Future Inheritance: Choice or Chance?*, Oxford University Press, London, 1974

Karp, L. E. *Genetic Engineering: Threat or Promise?*, Nelson-Hall, Inc., Chicago, 1976

Leach, G. *The Biocrats*, Jonathan Cape Ltd., London, 1970

Luria, S. E. *Life: The Unfinished Experiment*, Charles Scribner's Sons, New York, 1973

Milunsky, A. and G. J. Annas. *Genetics and the Law*, Plenum Press, New York, 1976

Nilsson, L., A. Ingelman-Sundberg, and C. Wirsén. *A Child Is Born*, Dell Publishing Co., Inc., New York, 1969

Paoletti, R. A. (editor). *Selected Readings: Genetic Engineering and Bioethics*, 2nd edition, MSS Corporation, New York, 1974

Pines, M. *The Brain-Changers: Science and the New Mind Control*, Harcourt, Brace, Jovanovich Inc., New York, 1973

Quinlan, J. and J. Quinlan, with P. Battelle. *Karen Ann*, Doubleday & Company, Inc., New York, 1977

Ramsey, P. *Ethics at the Edges of Life: Medical and Legal Intersections: The Rampton Lectures in America*, Yale University Press, New Haven, Conn., 1978

Ramsey, P. *Fabricated Man*, Yale University Press, New Haven, Conn., 1970

Rogers, M. *Biohazard*, Alfred A. Knopf, New York, 1977

Rorvik, D. and L. B. Shettles. *Choose Your Baby's Sex*, Dodd, Mead & Company, New York, 1977

Rosenfeld, A. *Prolongevity*, Alfred A. Knopf, New York, 1976

Rosenfeld, A. *The Second Genesis: The Coming Control of Life*. Prentice-Hall, Inc., Englewood Cliffs, N.J., 1969

Sagan, C. *The Dragons of Eden: Speculations of the Evolution of Human Intelligence*, Random House, Inc., New York, 1977

Stevens, L. A. *Explorers of the Brain*, Alfred A. Knopf, New York, 1971

Taylor, G. R. *The Biological Time Bomb*, New American Library, New York, 1969

Veatch, R. M. *Case Studies in Medical Ethics*, Harvard University Press, Cambridge, Mass., 1977

Wade, N. *The Ultimate Experiment: Man-Made Evolution*, Walker and Company, New York, 1977

Williams, G. *The Sanctity of Life and the Criminal Law*, Alfred A. Knopf, New York, 1970

Wolstenholm, G. E. W. and D. W. Fitzsimons (editors). *Law and Ethics of A.I.D. and Embryo Transfer*. Ciba Foundation Symposium 17 (new series), Associated Scientific Publishers, Amsterdam, 1973

Young, J. Z. *An Introduction to the Study of Man*, Oxford University Press, London, 1971

ARTICLES

"Can aging be cured?" *Newsweek*, pp. 56–66 (Apr. 16, 1973)

Clark, M. with M. Gosnell and D. Shapiro. "The new war on pain," *Newsweek*, pp. 48–58 (Apr. 25, 1977)

Crowe, J. H. and A. F. Cooper, Jr. "Cryptobiosis," *Scientific American*, 225(6): 30–36 (1971)

Duff, S. and A. G. M. Campbell. "Moral and ethical dilemmas in the special-care nursery," *New England Journal of Medicine*, 289: 890–894 (1973)

Epstein, C. J. and M. S. Golbus. "Prenatal diagnosis of genetic diseases," *American Scientist*, 65: 703–711 (1977)

Etzioni, A. "Sex control, science, and society," *Science*, 161: 1107–1112 (1968)

"Exploring the frontiers of the mind," *Time*, pp. 50–59 (Jan. 14, 1974)

Friedmann, T. and R. Roblin. "Gene therapy for human genetic disease?" *Science*, 175: 949–955 (1972)

Gore, R. "The awesome worlds within a cell," *National Geographic*, pp. 355-395 (Sept., 1976)

Grossman, E. "The obsolescent mother: a scenario," *Atlantic*, 227: 39-50 (1971)

Gwynne, P. with T. Clifton, M. Hager, S. Begley, and R. Gastel. "All about that baby," *Newsweek*, pp. 66-72 (Aug. 7, 1978)

Karp, L. E. and R. P. Donahue. "Preimplantation ectogenesis—science and speculation concerning *in vitro* fertilization and related procedures," *Western Journal of Medicine*, 124: 282-298 (1976)

Kass, L. R. "Babies by means of in vitro fertilization: unethical experiments on the unborn?" *New England Journal of Medicine*, 285: 1174-1179 (1971)

Kass, L. R. "Making babies—the new biology and the 'old' morality," *Public Interest*, pp. 18-56 (Winter, 1972)

Kennedy, I. "The legal effect of requests by the terminally ill and aged not to receive further treatment from doctors" *Criminal Law Review*, pp. 217-232 (Apr., 1976)

"A life in the balance," *Time*, pp. 52-61 (Nov. 3, 1975)

"Man into superman: the promise and peril of the new genetics," *Time*, pp. 33-52 (Apr. 19, 1971)

McCormick, R. A. "To save or let die—the dilemma of modern medicine," *Journal of the American Medical Association*, 229: 172-176 (1974)

"The modern men of parts," *Time*, pp. 73-75 (Mar. 18, 1974)

Peckins, D. M. "Artificial insemination and the law," *Journal of Legal Medicine*, 4: 17-22 (1976)

"Recombinant DNA research: a debate on the benefits and risks," *Chemical & Engineering News*, pp. 26-42 (May 30, 1977)

"The sale of human body parts," *Michigan Law Review*, 72: 1182-1264 (1974)

Sanders, H. J. "Artificial organs," *Chemical & Engineering News*, pp. 32-49 (Apr. 5, 1971) and pp. 68-76 (Apr. 12, 1971)

Seligmann, J. with M. Gosnell and D. Shapiro. "New science of birth," *Newsweek*, pp. 55-60 (Nov. 15, 1976)

Sinsheimer, R. "An evolutionary perspective for genetic engineering," *New Scientist*, 73: 150-152 (1977)

Snow, C. P. "Human care," *Journal of the American Medical Association*, 225: 617-621 (1973)

"Tinkering with life," *Time*, pp. 46-51 (Apr. 18, 1977)

Villet, B. "Opiates of the mind," *Atlantic*, pp. 82-89 (June, 1978)

Watson, J. D. "Moving toward the clonal man: is this what we want?" *Atlantic*, 227: 50-53 (1971)

Westoff, C. F. and R. R. Rindfuss. "Sex preselection in the United States: some implications," *Science*, 184: 633-636 (1974)

Wiley, J. P., Jr. and J. K. Sherman. "Immortality and the freezing of human bodies," *Natural History*, 80 (10): 12-22 (1971)

Index